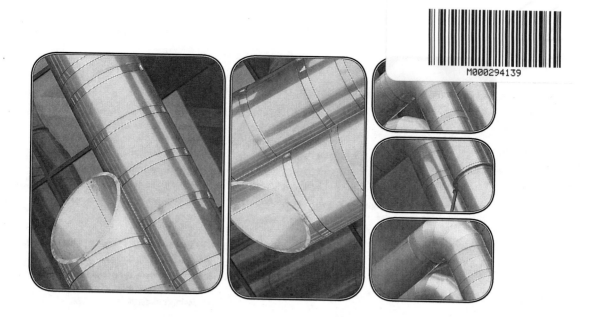

Safety, Health, and Environmental Concepts for the Process Industry

Michael Speegle

Second Edition

DELMAR
CENGAGE Learning™

Australia • Brazil • Japan • Korea • Mexico • Singapore • Spain • United Kingdom • United States

**Safety, Health, and Environmental Concepts
for the Process Industry, Second Edition**
Michael Speegle

Vice President, Editorial: Dave Garza

Director of Learning Solutions: Sandy Clark

Executive Editor: David Boelio

Associate Acquisitions Editor: Nicole Sgueglia

Managing Editor: Larry Main

Senior Product Manager: Sharon Chambliss

Editorial Assistant: Courtney Troeger

Vice President, Marketing: Jennifer Baker

Marketing Director: Deborah Yarnell

Associate Marketing Manager: Jillian Borden

Production Director: Wendy Troeger

Production Manager: Mark Bernard

Content Project Management: PreMediaGlobal

Art Director: Joy Kocsis

Technology Project Manager: Christopher
Catalina

Production Technology Analyst: Joe Pliss

Compositor: PreMediaGlobal

For product information and technology assistance, contact us at
Cengage Learning Customer & Sales Support, 1-800-354-9706
For permission to use material from this text or product,
submit all requests online at **www.cengage.com/permissions**.
Further permissions questions can be e-mailed to
permissionrequest@cengage.com

Library of Congress Control Number: 2011942152

ISBN-13: 978-1-133-01347-1

ISBN-10: 1-133-01347-3

Delmar
Executive Woods
5 Maxwell Drive
Clifton Park, NY 12065
USA

Cengage Learning is a leading provider of customized learning solutions with office locations around the globe, including Singapore, the United Kingdom, Australia, Mexico, Brazil, and Japan. Locate your local office at:
www.cengage.com/global

Cengage Learning products are represented in Canada by Nelson Education, Ltd.

To learn more about Delmar, visit **www.cengage.com/delmar**

Purchase any of our products at your local college store or at our preferred online store **www.cengagebrain.com**

Notice to the Reader
Publisher does not warrant or guarantee any of the products described herein or perform any independent analysis in connection with any of the product information contained herein. Publisher does not assume, and expressly disclaims, any obligation to obtain and include information other than that provided to it by the manufacturer. The reader is expressly warned to consider and adopt all safety precautions that might be indicated by the activities described herein and to avoid all potential hazards. By following the instructions contained herein, the reader willingly assumes all risks in connection with such instructions. The publisher makes no representations or warranties of any kind, including but not limited to, the warranties of fitness for particular purpose or merchantability, nor are any such representations implied with respect to the material set forth herein, and the publisher takes no responsibility with respect to such material. The publisher shall not be liable for any special, consequential, or exemplary damages resulting, in whole or part, from the readers' use of, or reliance upon, this material.

Printed in the United States of America
2 3 4 5 6   21 20 19 18 17

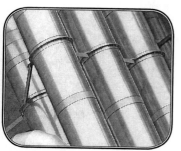

Contents

Contents

Contents

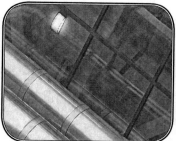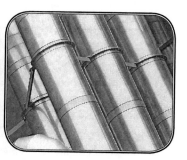

Preface

I worked in the process industry for 18 years, almost half of that time as a technician and the other half as a supervisor. I was well trained in the mandated safety and environmental subjects, and as a supervisor, taught those subjects to my technicians and enforced site safety rules. I confess that I never did comprehend the enormous amount of knowledge and economics involved in an entire plant's safety, health, and environmental (SHE) infrastructure and administration. It wasn't until I began writing this book that I received a good education in SHE. I think that is true for the majority of us who work in the processing industry—we can't see the forest for the trees. A refinery or chemical plant is a large complex of many systems and most of us only comprehend a small subset of those systems. We have only a vague knowledge of the embedded safety, health, and environmental "systems" within the site's systems.

When I began to teach safety, health, and environment in the process technology curriculum I was disappointed with the lack of a good textbook written at a level for the process technician/student. I found several fine textbooks written for industrial hygienists and occupational safety managers, but my intended audience would have to struggle to plow through those texts and a significant amount of the material was not applicable. Since I am both a fiction and nonfiction writer, I decided to write the kind of textbook I needed for my students and that might also be utilized by industry. That's when my real SHE education began.

This book has been written specifically for the process industries—industries such as refining, petrochemicals, electric power generation, food processing and canning, and paper mills. Millions of dollars have been invested, and continue to be invested, in the processing industries. The investment is made with a belief and commitment to a return on that investment; in other words, a profit. If a company is producing the right product at the right time with little competition, profits come easily. However, when a company has a lot of competition, as do refineries and chemical plants, plus strict health, safety, and environmental constraints, profits do not come easily. Companies strive to protect their investments in people, equipment, and potential profits from fires, explosion, and expensive litigation. Hence, the need for good training and training materials, which I hope this textbook fulfills as a function of a good training resource.

Based on the suggestions from colleagues at other colleges, I revised my order of chapters. This edition starts with the chapters on the technician responsibilities, history of accidents, and types and frequency of accidents. Next follows chapters of the various hazards in the processing and manufacturing industry. These chapters inform the reader of numerous bad things that can happen, followed by chapters that reveal how industry has planned and implemented to prevent those bad things from happening. The reader is exposed to numerous federal standards and their enforcement and the reader realizes there are severe penalties if a site fails to protect its workforce and surrounding community. The text ends with the same chapter as the first edition, the one on hurricanes and plant security. Why include a chapter on hurricanes? Hurricanes are a real hazard to the plants in the Gulf Coast area where a large number of petrochemical and refining plants are located. They are a hazard prepared for on a frequent basis.

I have also mentioned instrumentation and analyzer technicians in this text and included several photographs since they are out there on the units exposed to the same hazards as a process technician. This was suggested to me by colleagues at a nearby college.

To recap what is new to this second edition:

1. Revised order of chapters.
2. Combining of two chapters with the elimination of material considered not needed based on colleagues comments.
3. Added information on PPE.
4. Inclusion of analyzer and instrumentation technicians in the text.
5. Addition of a large number of photographs.
6. Large increase in information about environmental compliance, various standards, including Title V, and how analyzers assist with compliance.
7. A *Resource* section at the end of each chapter that guides students to Internet resources on the various subjects of the chapter.

It is a fact that the process technician with his or her intimate knowledge of the manufacturing process can contribute substantially to the bottom line of a company. This book was written specifically for the process technician with the intent of making him or her a significant contributor to a company's competitive edge by reducing accidents and injuries. This book is a general overview of safety, health, and environmental issues that affect the processing industries and meets almost all of the course objectives for a safety, health, and environment curriculum as determined by the Gulf Coast Process Technology Alliance.

Mike Speegle

CHAPTER 1

The Process Employee's Role in Safety, Health, and Environment

Learning Objectives

Upon completion of this chapter, the student should be able to:

- *Describe how the process employee's role has changed in the last 30 years.*

- *Discuss the importance of employee safety to the process industries.*

- *List the roles of today's process employees in safety, health, and the environment.*

- *Explain what has caused the roles of process employees to change from what they were 30 years ago.*

- *Explain why all risk cannot be removed from process industry jobs.*

INTRODUCTION

In this chapter, we will discuss the role of the process employee in the processing industries. *Processing industry* is broad term that most simply means taking a raw material and converting (processing) it into a valuable product for sale. Process employees who work the processing areas of a site are its operators, instrumentation and analyzer technicians, and maintenance personnel. All are responsible during the normal course of their duties to remain in compliance with company, federal, state and local safety, health, and environmental (SH&E) regulations. However, it should be noted that quite a few of the subjects in this textbook apply to other occupations, such as manufacturing, mining, oil and gas exploration, and production, to name a few. Quite a few occupations require respirator protection, personal protective equipment (PPE), hazard recognition, etc. In this textbook, we will also discuss employee roles in new technologies and how these roles impact safety, health, and the environment.

Incidentally, *safety* and *health*, although closely related are not the same thing. Safety is often thought of as being concerned with injury-causing situations or with hazards to humans that result from sudden severe conditions. Health is often thought of as being concerned with disease causing conditions or prolonged exposure to dangerous but less intense hazards. However, the line between safety and health is not always clearly drawn. As an example, stress can cause health problems through adversely affecting a worker's physiology or psychology. However, an over-stressed worker might be more prone to accidents and forgetful of safety precautions. The point being made is there is a difference between the concepts of *safety* and *health* though certain conditions may blur the differences.

Concern for the safety of personnel and equipment is vital to employees and management alike. Each has a personal stake in safety. The company (management) is concerned about safety because it is (1) an ethical responsibility and (2) it affects the bottom line (profits). Most companies have safety programs that deal with day-to-day precautions that must be taken while performing work. There are also mandates from federal, state, and local regulatory agencies that require periodic training and testing on specific safety and environmental regulations. Failure to do this training may result in expensive accidents and penalties. Employees are concerned because, after long years of service, they would like to retire healthy and vigorous enough to enjoy their retirement years.

RISK IN PROCESSING INDUSTRIES

The manufacturing facility where a person works (see Figure 1-1) can be as safe as a baby's padded crib or as hazardous as a rattlesnake den, depending upon how it is operated and maintained, and its culture of safety. Depending on the size of the process unit there will be hundreds, if not thousands, of miles of piping, hundreds of valves and flanges, high temperatures, toxic and/or flammable chemicals, plus heights and noise. Employees are taught the safety rules and the consequences for failing to observe them. *Safety is an attitude* and that attitude is manifested in the employees planning their work to include protective equipment and remaining alert for the unexpected. It is the job of the employee to avoid accidents and prevent down time that can result from not following established safety rules. Careless attitudes result in careless workers.

Here is an important statement to remember: **There is no such thing as a risk-free environment.** When we go to bed at night, we expect to wake up in the morning, but we hear of people dying in their sleep. When we get in our car for a trip to the mall, we expect to get there and back safely, but the evening news gives daily reports of fatal traffic accidents despite traffic laws, seat belts, and inflatable airbags. Risk is everywhere (see Figure 1-2).

Working in a processing industry is a lot like playing in a football game. Football players enter the game wearing protective equipment and trained to block and tackle. They

Figure 1-1 Process Unit

Figure 1-2 Risk Is Everywhere

exercise vigorously to be in shape to prevent or minimize injuries during the game. The rules of the game have been explained to them and they understand them. They know the team will suffer a penalty if they break the rules. In fact, their infraction may cost the team the game. In the same manner, a process employee can view their processing site as a playing field, themselves as a player, and safety, health, and environmental rules are the game rules.

Today, the process industry is obligated through goodwill, ethics, and law to keep the surrounding community informed of the risks and the corrective measures it takes to manage those risks. There is a trade-off between reasonable safety measures and excessive safety measures. If the public demands tires, plastics, nylon clothing, gasoline, jet fuel, or other things made from hydrocarbons and chemicals, it will have to accept the risk that comes with having process industries. Process safety decisions are often risk-based because they concern issues that are not often solved by simple rules or covered by existing codes or regulations. A risk-based decision inherently includes economics as one consideration. Corporations want to build and operate plants in a manner that will protect the health and safety of employees, processing equipment, and the surrounding public. Yet, they cannot eliminate all risk. They can only control or reduce it to an acceptable level based on sound design, engineering controls, and regulatory requirements. Once that level of safety is achieved, it becomes prudent to consider economics in deciding whether to pursue additional risk avoidance expenditures. It is possible to design a risk-free plant at a cost that will make it prohibitive to build or operate.

In a like manner, the process employee working inside a refinery that converts crude oil into gasoline and jet fuel must accept the risk that comes with their job. They may not like working around tens of thousands of gallons of highly flammable material, extreme heat, high pressures, and toxic chemicals on a daily basis, but if they adhere to their safety policies and training their risk is minimal. They are in greater danger of being killed in an auto accident then being killed at work in a refining or petrochemical plant. The processing industry has learned to manage risk. Governmental agencies have issued many regulations—hazard communication, hazardous waste operations, and process safety management—designed to address the issue of risk and its management. These will be covered in later chapters.

THE PROCESS EMPLOYEE'S ROLE

In response to the changing industrial scene, process employees have assumed a multifaceted role. Today's employees no longer confine themselves to just production work; they are active in safety committees, health issues, public relations, and environmental concerns. However, management now regards safety, health, and environmental responsibility a vital part of all employee's jobs (see Figure 1-3), shared with production and quality goals and responsibilities.

When new safety or environmental regulations are issued by regulatory agencies, the regulations that apply to a plant site are first reviewed by the SH&E group. Let us pretend that Occupational Safety and Health Administration (OSHA) just passed a new standard (law). The SH&E department, with assistance from operations and maintenance, would

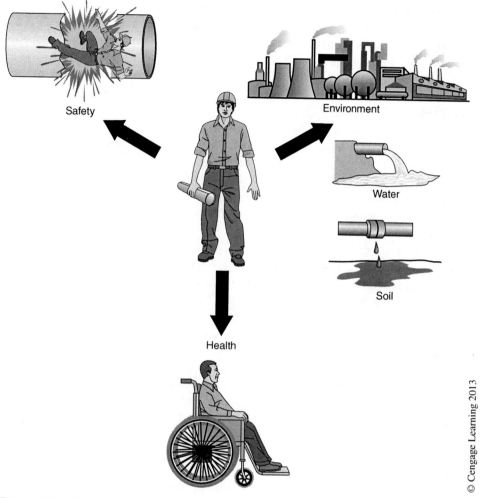

Safety

Environment

Water

Soil

Health

Figure 1-3   Employee Role in Safety, Health, and the Environment

study the standard and determine how it applies to its site. A typical sequence of events might be:

- Help develop and proof procedures which are site specific
- Pilot the procedures and make changes where needed
- Obtain any additional equipment or tools not possessed by the production site
- Conduct training sessions for affected employees
- After implementation date, enforce the procedures

As an example of how all of this comes together, let's look at some procedures mandated either by plant policy or governmental regulation. Suppose an employee has to have a pump worked on. This would involve writing a work permit, lockout and tagout of the pump, and draining and purging of the pump to make it ready for the maintenance personnel. Safety and health employees do not do this, operations employees do this. As another example, during turnarounds, employees are responsible for draining equipment, developing blind lists, work-orders, fire watch, and the monitoring of appropriate use of safety equipment and procedures by contractors working on their unit. SH&E personnel do not do the tasks described above. *Process operators do them.* The SH&E personnel are merely available as a resource and for guidance.

THE MANAGEMENT OF SAFETY

Today, a highly skilled and trained work force is a competitive necessity in a fast changing technological society. The pace of technological developments dictates that an organization evolves with changes in technology. An organization cannot afford to remain static and survive. Change requires training on new equipment, modified processes, and new regulatory standards. Plus, management is always on the look out for more advanced and better equipment, training systems, and safety software.

Improvements in productivity resulting from digitized record keeping are a significant help to companies. One key benefit is that important safety related information is available quickly in specific categories and formats. These systems can answer a question like: "How many years out of the last five years did we have an OSHA incident rate below 5.0?" Today organizations recognize that digitized storage of data is a gold mine of information that can direct a company to a better safety record and protect the company from malicious or frivolous lawsuits.

The management of safety is now treated like another branch of business management. Historically, safety had been regarded as a low-tech job and viewed as no more challenging than buying a few safety shoes, safety glasses, and displaying some posters and slogans. In today's business environment, it has become clear that safety must be managed like any other important business function. Safety activities require expenditures. These expenditures go toward ensuring a safe (or safer) work place. Managers recognize that a safe worker is productive worker. Better safety performance translates into real dollar savings in terms of lower insurance premiums and lower workman compensation costs. Workers who feel safe and healthy are less likely to make mistakes and have better morale. Safety managers today quantify not only the costs but the savings and benefits of a safe workplace. It is much cheaper to strive for a safe workplace. Trust, openness, and a spirit of cooperation between management and workers can develop a dedicated workforce. In fact, the distinction between management and labor may be an impediment to safety since management should be considered just another part of the workforce.

THE PROCESS EMPLOYEE AND CHANGE

Let's do a quick review of some major changes that have occurred in the processing industry over the last 20 to 30 years. The Gulf Coast Process Technology Alliance (GCPTA), an alliance created in 1997 of industry representatives, trade associations, and educational institutions, has helped identify some of the changes. They are:

- A much more diverse workforce
- Extensive computerized controls and automation
- Workers must understand and comply with safety and environmental regulations

- Workers are integrated into teams and must have interpersonal skills
- Workers must support and contribute to process quality and a quality improvement process
- Workers are involved with process **hazard analyses**

Because of the sophistication of new automated control systems and the complexity of the growing list of responsibilities, the requirements for a process employee changed. No longer can they be someone who could come in off the street without training or experience and be trained for the job in a few days or even a few weeks. The fact that operators, instrumentation technicians, and analyzer technicians are now more frequently referred to as technicians implies a change in their roles. The definition of *technical* is: *having special or practical knowledge of a mechanical or scientific subject.*

Process employees today are required to have *special or practical knowledge and skills of a mechanical or scientific nature* in addition to interpersonal skills. They are being hired for *knowledge.* Part of the reason for that is because of their increased responsibilities involving technical equipment, safety, and the environment. Since plants are highly automated, brute strength is no longer a requirement of an employee. Plants now seek employees who can think, analyze, solve problems, and respond in correct ways with minimal supervision. Some of these problems will be safety or environmental problems. The old days of a supervisor doing the thinking for everyone and telling everyone exactly what to do and solving all problems are gone. Companies hire employees capable of critical thought and that have some college math, analytical skills, and communication skills. Employees also need to be familiar with computers and some computer programs. They must understand the economics of their process. They will be required to continually improve their knowledge and skills because changing technology will require them to continue learning and schooling. After receiving training, process employees are responsible for running their unit economically, safely, and efficiently.

Process control systems have increased in complexity as electronic control systems were replaced by computer-directed distributed control systems (DCS). Gone are the control rooms full of panels and gauges, switches, meters, and charts. They have been replaced with something that looks like a space shuttle control panel. Processes run faster, more safely, and produce higher quality product using the latest methods of statistical quality control. Improvements in information technology make it possible for the employees to know almost immediately what each piece of equipment under their control is doing.

Process Employees

Who are the process employees responsible for the safety, health, and environmental compliance of their site that this book is addressing? They are the process technician (see Figure 1-4), the instrumentation technician (see Figure 1-5), and analyzer technician (see Figure 1-6). They are the personnel who spend almost their entire day on the processing unit among the vessels, piping, pumps, and equipment. They are the employees responsible for controlling the unit, detecting problems and fixing them, providing reliable data to the control room, and repairing broken equipment. This is no slight at the engineers who are assigned to the process unit; however, they spend the majority of their time away from the unit working on numerous administrative assignments and tasks in their offices.

Figure 1-4   Process Technician

Figure 1-5   Instrument Technician

© Cengage Learning 2013

Figure 1-6 Analyzer Technician

The process, instrument, and analyzer technicians are all exposed to the health and physical hazards of the unit each minute they are among the noisy vibrating equipment. All have to climb significant heights to taking readings and collect samples, replace broken instruments, and calibrate instruments and analyzers. They all must use the permit system and consult Material Safety Data Sheets (MSDS) because of the chemicals they will encounter doing their jobs. They are all exposed to the hazards of the chemicals of the process, the high temperatures and pressures of the process, and the hazards associated with rotating or moving equipment. All, working as a team, help achieve safety goals and environmental compliance.

Employees and the Environment

Unlike operators 20 or 30 years ago, process employees are often required to assist SH&E personnel with air, water, and solid waste regulations, plus safety regulations. Employee environmental responsibilities are covered in Chapter 20. It cannot be emphasized enough how important environmental compliance is to a processing site, not just because of the expensive fines and penalties but also because of the responsibility of being a good neighbor to the surrounding community.

New Roles and Responsibilities

Earlier, we said the petrochemical and refining industries were largely responsible for changing the process employee's job from one that required a manual laborer to one that required a skilled employee. Historically, the batch distillation units that produced only kerosene were replaced by more complex batch units that produced everything from fuel gases to heavy tars. These more complex batch units gave way to continuous processes which were mostly manually controlled units producing the simpler petrochemical derivatives. The continuous flow process could operate around the clock and around the calendar. This introduced shift work and rotating shift schedules for operators. This in turn injected a new safety and health problem to the industry because shift work reduces alertness and mental agility and affects worker health.

As more complex operations, such as catalytic cracking and reforming were introduced, more complex instrumentation and controls were needed to operate the plants safely and economically. Pneumatic units replaced manual controllers as operations became more complex. These were subsequently replaced by electronic controllers, then by programmed smart controllers, and eventually by computerized systems. Each evolution required that operators, instrument technicians, and analyzer technicians have more and more technical training to understand process operations and perform their job safely and profitably. A gradual evolutionary process occurred and many employees evolved into technicians. They were no longer just blue collar workers.

The increasing technical and regulatory environment requires a process employee to possess varied skills. Some of the more important ones are:

- **Technical expertise**—Process employees today must possess technical expertise. In the past, they were not expected to contribute to process improvements, be involved in quality, interact with customers, be aware of environmental issues, or be involved in visits by governmental agencies. They are now. Their value to the company is in terms of technical knowledge and skills, plus interpersonal skills.
- **Regulatory knowledge**—With new requirements, laws, and regulations being enacted, the employee must be aware of these changes and adhere to them in their daily work. In the past, a small chemical spill may not have been considered a serious concern. Today, process employees must document spills, classify them, and report them to the proper agencies. Failure to do so can result in the employee or employee's company being fined, someone being imprisoned, or both. With OSHA and the Environmental Protection Administration (EPA) taking active roles in plant safety, health, and environmental matters, most plants now have extensive programs to help meet compliance regulations.
- **Communication skills**—Lack of communication and poor communication are constant complaints in all businesses. Process employees must communicate effectively with fellow team members and other plant personnel. They should be capable of good verbal and written communication skills. Information should be clear, concise, and easily understood so that it can be acted on without error. The written report employee's prepare at the end of their shift should summarize their activities in a way that others can easily understand. The procedures and guidelines that they help write should be written so that misunderstandings and mistakes are eliminated. Many employee reports and logbooks are legal documents that can be referred to in case of accidents or audits by regulatory agencies.
- **Computer literacy**—The employee must also be very familiar with computers and several types of computer programs. The computer is an

important tool for issuing maintenance work orders, for tabulating records and data from the unit and the laboratory, and for maintaining personnel records such as timesheets, payroll, and vacation schedules. Much of an employee's training will take place via computers. And more importantly, employees must understand the control schemes for their units. Plants today have much of their operations controlled by computers using sophisticated DCS for this purpose.

- **Interpersonal skills**—The process employee must have good interpersonal skills that allow them to work as an effective team member. Each crew must function as a team to do its job effectively. The first team an operator becomes assigned to is the crew they will work with on the processing unit. They will be with that crew for 25–30 years until they retire. They are going to have to get along with everyone on the team. The operations and support groups must work together as a team to resolve process or equipment problems. Employees may also be asked to work on special teams to troubleshoot or upgrade equipment, to review safety or environmental issues, or to write operating or maintenance procedures. Interpersonal skills and teamwork is an important part of the employee's job today.

- **Problem solver**—Gone are the days when something went wrong on a processing unit and engineers and/or first line supervisors took care of the problem. The role of problem solver is one on the most important that the process employee must assume. They must be able to troubleshoot problems in operating equipment, instruments and analyzers and identify the problem. They must become so familiar with their unit or equipment that they can quickly recognize when an abnormal situation occurs. Operators must know how to adjust unit conditions to correct for quality and yield loss problems when they occur. They must recognize hazardous conditions that require corrective action.

- **Trainer**—The process employee must use all the aforementioned skills to help train new process employees in their roles and responsibilities. This includes not just process training, but also safety and environmental training. Much of the training involving a specific process or unit is done one-on-one using experienced employees.

- **Quality and continuous improvement**—Quality and continuous improvement are requirements for survival in today's highly competitive markets. In recent years, continuous improvement has become a relentless goal for all organizations. The process employee must know every valve, pipe, vessel, and the in-and-outs of their unit, instrumentation or analyzers better than anyone else. They are the most qualified for defining the large and small pathways that lead to continuous incremental improvements and higher profitability.

SUMMARY

Process safety decisions are often risk-based because they concern issues that are not often controlled by simple rules or covered by existing codes or regulations. A risk-based decision inherently includes economics as one consideration. The processing industry has learned to manage risk.

One of the most important roles of a process employee is to carry out safety, health, and environmental (SH&E) functions. Historically, safety had been regarded as a minor priority, but in today's business environment, safety must be managed just like any other business function. Managers recognize that a safe worker is productive worker. Better safety records translate into real dollar savings in terms of lower insurance premiums and lower workman compensation costs.

Process employees today are required to have *special or practical knowledge and skills of a mechanical or scientific nature.* Employees today must have skills such as (1) technical expertise, (2) communication skills, (3) troubleshooting abilities, and (4) computer literacy.

REVIEW QUESTIONS

1. The three process employees responsible for safety and environmental compliance on a processing unit are the _____, _____, and _____.

2. List five roles of a process employee today.

3. (T/F) Safety is an attitude.

4. Explain the importance of those five roles to the process industry.

5. Explain what is meant by "reasonable risk."

6. (Choose the two best) Better safety performance translates into dollar savings by:
   a. Lower site insurance premiums
   b. Reduced firefighting
   c. Lower workman compensation costs
   d. Less fines from government agencies

7. Explain why working in the processing industry is a lot like playing in a football game.

8. Two reasons environmental compliance is important to a processing site are _____ and _____.

9. Three reasons for an employee to be computer literate are _____, _____, and _____.

10. _____ are important because you will be assigned to a crew that you will have to get along with for 20 or more years.

EXERCISES

1. Write a one page report describing why process operators, instrument technicians, or analyzer technicians are no longer considered blue-collar workers.

2. Write a one page report describing the risks you take when you take the following vacation:
   a. Drive from your home on the freeway to the airport.
   b. Fly to Las Vegas.
   c. Rent a car and drive to the Grand Canyon.
   d. Raft the Grand Canyon, spending two nights sleeping on the river bank before the rafting is over.
   e. Fly back home.

CHAPTER 2

History of the Safety and Health Movement

Learning Objectives

Upon completion of this chapter the student should be able to:

- *List three reasons for improvements in industrial safety.*

- *List four important events in the history of the safety movement after 1900.*

- *Discuss how* settlement houses *played a part in occupational safety.*

- *Discuss organized labor's part in the safety movement.*

- *List five occupational diseases and their causes.*

- *List the Three E's of safety and explain the function of each.*

- *Identify three important safety organizations and explain their roles in safety.*

INTRODUCTION

Safety and health awareness has a surprising history that dates as far back as the time of the Egyptian pharaohs. The Code of Hammurabi, named after a Babylonian king circa 2000 BC, contained clauses that could be interpreted as early attempts at workers' compensation. There is also evidence of concern for safety and health during the time of the Roman Empire. This chapter examines the history of the safety movement in the United States and how it has developed over the years.

In America in the early 1900s, industrial accidents were commonplace. In 1907, over 3,200 people died in mining accidents. During this period, legislation, precedent, and public opinion all favored management. If workers were injured, it was due to their own carelessness. Injured workers received no compensation and were usually fired for being so careless as to get injured. No matter if it was the company's fault. Few considerations were given to a worker's safety. On the job, it was literally every man for himself. Working conditions for industrial employees today have vastly improved. According to the National Safety Council (NSC), the current death rate from work-related injuries is approximately four per 100,000 or less than a third of the rate 50 years ago. Safety improvements up to now have been the result of:

- Pressure for legislation to promote safety and health
- Costs associated with accidents and injuries
- Recognition that safety and health concerns rank in importance with production and quality

Improvements in safety and health in the future are likely to come as a result of greater awareness of the cost effectiveness of a safe workplace and the competitive advantage gained from a safe and healthy workforce.

The NSC, after several permutations, was established in 1912, and plays a very important role in collecting data on accidents and injuries and making the information available. Today, the NSC is the largest organization in the United States devoted solely to safety and health practices and procedures. Its purpose is to prevent the losses arising from accidents or from exposure to unhealthy environments. Although chartered by an act of Congress, the NSC is a nongovernmental, not-for-profit, public service organization.

OVERVIEW OF THE SAFETY MOVEMENT IN THE UNITED STATES

The safety movement in the United States can trace its roots to England. During the Industrial Revolution, child labor in factories was common and children as young as six years old worked long hours, often in unhealthy and unsafe conditions. After an outbreak of fever among the children working in cotton mills, the people of Manchester, England, began demanding better working conditions in the factories. In 1802, public pressure eventually forced a government response and the Health and Morals of Apprentices Act was passed in Great Britain. This legislation was significant because it marked the beginning of government involvement in workplace safety.

As in the rest of the newly industrialized nations, hazardous conditions were commonplace in the industrial sector of the United States. The safety movement began in America just after the Civil War. A chronology of some important events follows.

1867–1900

- Factory inspection was introduced in Massachusetts in 1867.
- In 1868, the first barrier safeguard was patented for moving equipment.
- In 1869, the Pennsylvania legislature passed a mine safety law requiring two exits from all mines.
- The Bureau of Labor Statistics (BLS) was established in 1869 to study industrial accidents and report pertinent information about those accidents.

- In 1877, the Massachusetts legislature passed a law requiring safeguards for hazardous machinery. In the same year the Employer's Liability Law was passed, establishing the potential for employer liability in workplace accidents.
- In 1892, the first recorded safety program was established in a Joliet, Illinois, steel plant in response to a scare caused when a flywheel exploded.
- Around 1900, Frederick Taylor began studying efficiency in manufacturing with the purpose of identifying the impact of various factors on efficiency, productivity, and profitability. Although safety was not a major focus of his work, Taylor did draw a connection between lost personnel time and management policies and procedures.

1901–Present

- In 1907, the U.S. Department of the Interior created the Bureau of Mines to investigate mining accidents, examine health hazards, and make recommendations for improvements.
- In 1908, an early form of workers' compensation was introduced in the United States.
- In 1913, the NSC was founded, the premier safety organization in the United States.
- From 1918 through the 1950s, safety awareness continued to grow and the federal government encouraged federal contractors to implement and maintain a safe work environment. At this time the government could only encourage manufacturers and businesses, not require nor penalize them.
- The 1960s saw the passage of legislation promoting workplace safety. The Service Contract Act of 1965, the Federal Metal and Non-metallic Mine Safety Act, the Federal Coal Mine and Safety Act, and the Contract Workers and Safety Standards Act all were passed during this decade.

The persistent and continued increases in death and injuries in industry were the primary reasons behind passage of the Occupational Safety and Health Act of 1970 (OSHA) and the Federal Mine Safety Act of 1977. These federal laws, particularly OSHA, represent the most significant legislation to date in the history of the safety movement. Table 2-1 summarizes some significant milestones in the development of the safety movement in the United States.

HISTORY OF OCCUPATIONAL HEALTH

Pathological conditions brought about by workplace conditions led to the development of a field of study called *occupational health*. The concept of occupational health as a support segment of manufacturing did not arise exclusively from the concern of the medical establishment.

In the later decades of the nineteenth century, millions of people emigrated from Europe to the United States and settled in its large cities. Steel mills, machine tool industries, railroads, canals, farms, and municipal governments provided employment for the new unskilled laborers with low wages. Almost all lived in poverty, had limited access to clean water, sewer systems, schools, or medical treatment. *Settlement houses* were an attempt to reverse these unanswered needs. Jane Addams founded Hull House in Chicago

Table 2-1 Safety Movement Milestones in the United States

1867	Massachusetts introduces factory inspection.
1868	Patent awarded for the first barrier safeguard.
1877	Massachusetts passes law requiring safeguards on hazardous machines and the Employer Liability Law is passed.
1892	First recorded safety program is established.
1900	Frederick Taylor conducts first studies of efficiency in manufacturing.
1907	Bureau of Mines created by the U.S. Department of the Interior.
1911	Wisconsin passes the first effective worker's compensation law.
1915	National Council on Industrial Safety formed (In 1900 it changed its names to the National Safety Council).
1916	Concept of product liability is established.
1970	Occupational Safety and Health Act passed.
1977	Federal Mine Safety Act passed.

© Cengage Learning 2013

to attempt to correct some of these social concerns. In response to medical needs, a public dispensary was established at Hull House, and a Visiting Nurses Association nurse made his headquarters there.

The settlement houses developed and became models that inspired some companies to initiate some occupational health and safety activities. In 1907, a group of cotton mills in the South employed trained nurses to attend persons who were sick in the mill community (to get them well as quickly as possible). At the New York Telephone Company, there was a "retiring-room" for operators who felt indisposed. The National Cash Register Company began giving physical examinations to applicants for work as early as 1901. Examinations were introduced at Sears, Roebuck, and Company in Chicago in 1909. By 1914, physical examinations of employees were a fixture in many large companies.

Organized Labor and Safety

Organized labor played a crucial role in the development of the safety movement in the United States. From the beginning of the Industrial Revolution in the United States organized labor has fought for safer working conditions and appropriate compensation for workers injured on the job. Many of the earliest developments in the safety movement were the result of organized labor's long and hard-fought battles against management's insensitivity to safety concerns.

Some of the most important contributions of organized labor to the safety movement were their work to overturn workplace anti-labor laws relating to safety. These laws were the *fellow servant rule*, the statutes defining *contributory negligence*, and the concept of *assumption of risk*. Each is briefly explained in the following bullets.

- The fellow servant rule held that employers were not liable for workplace injuries that resulted from the negligence of other employees. For example, if

Worker X tripped over a board and broke his leg because Worker Y neglected to remove the board, the employer was not liable.

- The doctrine of contributory negligence stated that if the actions of employees contributed to their own injuries, the employer was not held liable.
- The concept of assumption of risk held that people who accept a job assume the risks that go with it. Employees were not coerced into taking a job. Consequently, they should accept the consequences of their actions on the job rather than blaming the employer.

These were all employer-biased laws. Since the overwhelming majority of industrial accidents involve negligence on the part of one or more workers, employers rarely worried about liability. Since they could not be held liable, employers had little incentive to promote a safe work environment. Organized labor brought to the attention of the general public the deplorable working conditions employers allowed to exist. Public awareness and outrage eventually led to the employer-biased laws being overturned.

Occupational Diseases initiate Change

Lung disease in coal miners was a major problem in the 1800s, particularly in Great Britain, where much of the Western world's coal was mined. The lung disease, also known as the *black spit*, persisted from the early 1800s until around 1875 when it was finally eliminated by such safety and health measures as ventilation and decreased work hours. Miner lung diseases and compensation for diseased miners was debated in the United States until Congress finally passed the Coal Mine Health and Safety Act in 1969. The event that led to passage of this act was an explosion in a coal mine in West Virginia in 1968 that killed 78 miners. This tragedy focused attention on mining health and safety.

Over the years, the diseases suffered by miners were typically lung diseases caused by the inhalation of coal dust particulates. However, other miners developed a variety of diseases, the most common being silicosis. Once again, it took a tragic event—the Gauley Bridge disaster—to focus attention on a serious workplace problem. As part of a project to bring hydroelectric power to a remote part of West Virginia in the 1930s, a large tunnel had to be dug through the mountains. The tunnel cut through pure silica. Exposure to silica dust led to numerous deaths which provoked a public outcry. A fictitious account of the Gauley Bridge disaster entitled *Hawk's Nest* by Hubert Skidmore whipped public outcry into a frenzy, and in 1936, forced Congress to respond. Also in 1936, representatives from business, industry, and government attended the National Silicosis Conference, convened by the U.S. Secretary of Labor. Among other outcomes of this conference was a finding that silica dust particulates did cause silicosis.

Mercury poisoning is another health problem that has contributed to the evolution of the health and safety movement by focusing public attention on unsafe conditions in the workplace. In the early 1950s, a strange disease of the central nervous system appeared among the citizens of Minamata, a Japanese fishing village. Eventually there were 103 deaths, and some 700 persons were disabled. The disease was first noticed in the early 1930s. The disease with severe symptoms was common in Minamata, but extremely rare throughout the rest of Japan. After much investigation it was determined that a nearby chemical plant periodically dumped methyl mercury into the bay that was the village's primary source of food (fish and shellfish). The citizens of this small village ingested hazardous dosages of mercury every time they ate fish from the bay.

Mercury poisoning became an issue in the United States in New York City's hat-making industry in the early 1940s when it was noted that many workers in this industry displayed the same types of symptoms exhibited by the citizens of Minamata, Japan. This disease was often referred to as the "Mad Hatter's disease." Since mercury nitrate was used in the production of hats, a study was conducted and the study linked the symptoms of workers with the use of mercury nitrate. As a result, the use of this hazardous chemical in the hat-making industry ceased and a suitable substitute was found.

Another important substance in the evolution of the modern health and safety movement has been asbestos. At one time, asbestos was considered a wonder material and was widely used in the building construction industry. By the time it was determined that asbestos was hazardous and could cause lung cancer (mesothelioma), thousands of buildings contained the substance. As these buildings began to age, the asbestos began to break down and release dangerous microscopic fibers into the air. These fibers are so hazardous that removing asbestos from old buildings has become a highly specialized task requiring special equipment and training.

The effects of chromium compounds have been studied since World War II. While chromic acid mist was known to produce septal perforations in the nose, "chrome holes" of the skin, and chronic dermatitis, occupationally caused lung malignancies from certain chromate compounds were a newly discovered epidemiological finding. Exposure to vinyl chloride produced angiosarcoma (cancer of the liver) in workers exposed to the chemical. This wasn't discovered until an epidemiological survey revealed workers in plants that had vinyl chloride in the process had abnormally high incidences of liver cancer.

Since cancer of the scrotum in chimney sweeps was described in 1775, the incidence of work-related carcinogenesis (cancer-causing) has increased. One of the first to proclaim the relationship between certain industrial substances and cancer in workers exposed to toxic materials was Joseph Schereschewsky. In 1925, he completed a statistical review of the increase in cancer mortality over a 20-year period, and in 1927, advised the U.S. Public Health Service on a program of cancer research. Research in occupational cancer was among four recommendations submitted by a committee of investigators.

DEVELOPMENT OF ACCIDENT PREVENTION PROGRAMS

In the modern workplace, there are many different types of accident prevention programs (figure 2-1) ranging from the simple to the complex. Some of the most widely used accident prevention techniques include:

- Fail-safe designs
- Isolation
- Lockouts
- Screening
- Personal protective equipment

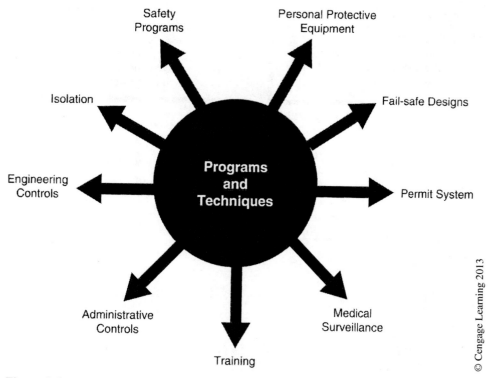

Figure 2-1   Development of Accident Prevention Techniques and Programs

These techniques are individual components of broader safety programs. Such programs have evolved since the late 1800s.

In the early 1800s, employers had little concern for the safety of workers and little incentive to improve unsafe work conditions. Organized safety programs were nonexistent, a situation that continued for many decades. During World War II, faced with manpower shortages, employers could not afford to lose workers to accidents. Industry began to realize the following:

- Improved engineering could prevent accidents
- Employees were willing to learn and accept safety rules
- Safety rules could be established and enforced
- Financial savings accrued from safety improvements

These realizations provided the incentive for employers to play an active role in creating and maintaining a safe workplace. This led to the development of organized safety programs sponsored by management. Early safety programs were based on the "Three E's of safety," (figure 2-2) which were *engineering, education*, and *enforcement*.

- The **engineering** aspects of a safety program involve making design improvements to both products and processes. Employers began to realize that processes used to manufacture products could be engineered to decrease potential hazards associated with the processes. Engineering controls remove the hazards or greatly reduce it so that serious harm cannot occur. Because of this, engineering controls is first in the hierarchy of hazard control.

21

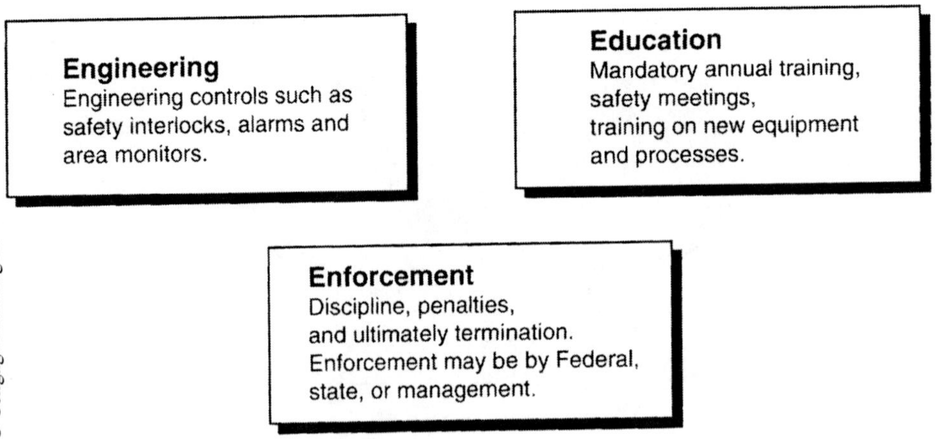

Engineering
Engineering controls such as
safety interlocks, alarms and
area monitors.

Education
Mandatory annual training,
safety meetings,
training on new equipment
and processes.

Enforcement
Discipline, penalties,
and ultimately termination.
Enforcement may be by Federal,
state, or management.

Figure 2-2   The Three Es of Safety

- The **education** aspect of a safety program ensures that employees know how to work safely, why it is important to do so, and that safety is a requirement of continued employment.
- The **enforcement** aspect of a safety program involves making sure that employees abide by safety policies, rules, regulations, practices, and procedures. Supervisors and fellow employees play a key role in the enforcement aspects of safety programs. If a rule is not enforced it is not a rule.

Important Safety Organizations

Today, numerous organizations are devoted in full or at least in part to the promotion of safety and health in the workplace (see Table 2-2). These lists are extensive now, but there was a time when these agencies and organizations didn't exist. We will discuss three important ones.

The grandfather *of* them all is the **National Safety Council**, mentioned earlier in this chapter. The NSC was founded in 1913 and charted by the U.S. Congress in 1953. It is the nation's leading advocate for safety and health. Its mission is to educate and influence society to adopt safety, health, and environmental policies, practices, and procedures that prevent and mitigate human suffering and economic losses arising from preventable causes.

The **Occupational Safety and Health Administration (OSHA)** is the federal government's administrative arm for the Occupational Safety and Health Act. Formed in 1970, some of OSHA's responsibilities are:

- Set and revoke safety and health standards
- Conduct inspections
- Investigate problems
- Issue citations and assess penalties
- Maintain a database of health and safety statistics

Table 2-2 Government Agencies Involved in Workplace Safety

Government Agencies Involved in Workplace Safety
American Public Health Association
Bureau of Labor Statistics
United States Consumer Product Safety Commission
National Institute for Standards and Technology
National Safety Council
Environmental Protection Agency
Occupational Safety and Health Administration
Bureau of National Affairs
National Institute of Occupational Safety and Health

© Cengage Learning 2013

Another governmental organization is the **National Institute of Occupational Safety and Health (NIOSH)**, which is part of the Centers for Disease Control of the Department of Health and Human Services. NIOSH is required to publish annually a comprehensive list of all known toxic substances. It will also provide onsite tests of potentially toxic substances so that employees know what they are handling and what precautions to take.

THE SAFETY AND HEALTH MOVEMENT TODAY

The safety and health movement has come a long way since the mid-1800s. It took a long time and paradigm shifts by both society and business management before safety and health was recognized as an important facet of every job and that it added to efficiency and profitability.

Today, there is widespread acceptance and understanding of the importance of providing a safe and healthy workplace. This understanding began during and after World War II when the various practitioners of occupational health and safety—safety engineers, safety managers, industrial hygienists, occupational health nurses, and physicians—began to see the need for cooperative efforts. Today, most businesses and manufacturing sites use an integrated approached for safety and health programs. This integrated approach is now the norm that typifies the health and safety movement today. By working together and drawing on their own respective areas of expertise, safety and health professionals are better able to identify, predict, control, and correct safety and health problems.

OSHA reinforces the integrated approach by requiring companies to have a plan for doing the following: (1) providing appropriate medical treatment for injured or ill workers; (2) regularly examining workers who are exposed to toxic substances; and (3) having a qualified first-aid person available during all working hours. Smaller companies may contract out these requirements; larger companies often maintain a staff of safety and health professionals. The health and safety staff in a modern industrial company might include the following positions:

- Safety engineers
- Safety managers
- Industrial hygienists
- Occupational health nurses
- Psychologists
- Physicians
- Emergency Response Personnel

New Safety and Health Problems

Maintaining a safe and healthy workplace is more complex than it has ever been. New materials and new processes have created new problems. About 8,000 new chemical compounds are created each year. Production materials have become increasingly complex and exotic. Engineering materials now include carbon steels, stainless steels, cast irons, tungsten, molybdenum, titanium, aluminum, powdered metals, plastics, etc. Each of these metals requires its own specialized processes and has its own associated hazards. Nonmetals are more numerous and have also become more complex. Plastics, plastic alloys, and blends, advanced composites, fibrous materials, elastomers, and ceramics also bring their own potential hazards to the workplace.

In addition to the more complex materials being used in modern industry and the new safety and health concerns associated with them, modern industrial processes are also becoming more complex and the potential hazards associated with them often increase. Here is a partial list of fairly new industrial equipment and processes, which may introduce new safety and health problems into the workplace.

- Computers
- Lasers
- Industrial robots
- Photochemical machining
- Laser beam machining
- Ultrasonic machining and chemical milling
- Expert systems
- Flexible manufacturing cells

New technologies, new materials, new equipment, and new chemicals will introduce new safety and health problems that will have to be detected, diagnosed, and eliminated or controlled. Process technicians working with new technologies and chemicals should be alert to hazards and quickly inform the appropriate personnel of the hazards.

SUMMARY

Milestones in the development of the safety movement in the United States include the following: first recorded safety program in 1892, creation of the Bureau of Mines in 1907, passage of the first effective workers' compensation law in the United States in 1911, and passage of OSHA in 1970. Organized labor has played a crucial role in the development of the safety movement in the United States. Particularly important was the work of unions to overturn anti-labor laws that inhibited safety in the workplace.

Specific health problems associated with the workplace have contributed to the development of the modem safety and health movement. These problems include lung diseases in miners, cancers caused by contact with various industrial chemicals, and lung cancer tied to asbestos. Widely used accident prevention techniques include failure minimization, fail-safe designs, isolation, lockouts, screening, personal protective equipment, redundancy, and timed replacements.

The development of the safety movement in the United States has been helped by the parallel development of safety organizations. Prominent among these are the NCS, the American Society of Safety Engineers, and the American Industrial Hygiene Association.

REVIEW QUESTIONS

1. Three reasons for improvements in industrial safety today are:
   a. Legislative pressure
   b. Required by the constitution
   c. Costs of accidents
   d. Recognition of the importance of safety and health

2. The purpose of the National Safety Council is to:
   a. Help protect the environment along with safety
   b. Prevent losses arising from accidents and unhealthy work environments
   c. Assess fines or penalties due to an unsafe workplace

3. (T/F) The safety movement in the America began just after the Civil War.

4. The _____ was established in 1869 to study industrial accidents and report information about those accidents.

5. Pathological workplace conditions led to the development of field of study called _____.

6. The _____ became models that inspired some companies to initiate some occupational health and safety activities.

7. The _____ rule held that employers were not liable for workplace injuries that resulted from the negligence of other employees.

8. The _____ stated that if actions of employees contributed to their own injuries the employer was not held liable.

9. _____ held that people who accept a job assume the risks that go with it.

10. Mercury was famous for causing two diseases, which were the _____ and _____ diseases.

11. The Three E's of safety are _____, _____, and _____.

12. The most important of the Three E's is _____.

13. Explain the function of each of the Three E's in safety.

14. _____ is required to publish annually a comprehensive list of all known toxic substances.

15. (T/F) OSHA requires a qualified first-aid person available during all working hours.

EXERCISES

1. About 8000 new chemicals—drugs, additives, plastics, etc.—are created each year. Google *NIOSH* and write a one page report on how this agency is involved in protecting workers and the public from hazards associated with these new chemicals.

2. Go to the Internet and research the Gauley Bridge disaster (discussed in this chapter) and write a two page report of this tragic event.

RESOURCES

www.wikipedia.org

www.osha.gov

CHAPTER 3

Accidents and Human Error

Learning Objectives

Upon completing this chapter the student should be able to:

- *List the benefits of accident investigations.*

- *List the five leading causes of accidental death in the United States.*

- *List the elements that make up the overall cost of an accident.*

- *List the five leading causes of work deaths in the United States.*

- *List four types of fatigue-producing designs.*

- *Explain four ways to minimize procedural errors.*

- *Explain the two-person concept.*

- *Describe three ways that workers can be involved in safety.*

INTRODUCTION

Unfortunately, fires, explosions, chemical leaks, and other incidents happen in the process industries. Depending upon the type of plant, its size, work force training, experience, and safety culture, a plant site may experience one or two incidents a year. An incident does not necessarily mean a serious fire or explosion. The most common types of incidents ranked by frequency are:

- Chemical leaks
- Fires

- Equipment failures
- Over-filled vessels

It is not uncommon in some industries, such as the chemical and electric utility industries, to experience new injuries or fatalities due to the same incidents. Industrial incidents have an eerie way of repeating themselves because organizations do not learn from the past. Individuals learn, but individuals retire or move to different locations and take their knowledge and experience with them. The organization as a whole loses memory. As people retire, move to other plants, or plants downsize, incidents of a similar type tend to recur within the same company at approximately 10-year intervals.

Investigations of industrial accidents reveal that most are caused by human error. The twentieth century's worst industrial disasters—Bhopal, Three Mile Island, and Chernobyl—helped clarify the complex chain of system problems that lead to human error. *System problems* are problems caused by a process system with built in design and operating deficiencies. The accidents provided numerous checklists and case studies for control room and equipment design. A partial list of some of the problems found at the industrial sites mentioned above are:

- Unprioritized alarm signals
- Malfunctioning equipment
- Poor maintenance practice
- Distant display and equipment control panels
- Inadequate operator training
- Poor communications
- Inadequate or outdated procedures

These accidents, and many less devastating incidents that continue to occur today, were linked by an incomplete analysis of human factors. The human side of safety was ignored with high costs. Chemical manufacturers have yet to get over the impact of Bhopal, which killed 3,800 and injured over 200,000. Litigation is still in process. It required 4.5 years and $970 million to clean up after the Three Mile Island nuclear plant accident. That cost is hundreds of millions more than the cost to build the plant. Long-term environmental and health impacts of Chernobyl continue to haunt Russia and her neighbors.

BENEFITS OF ACCIDENT INVESTIGATIONS

The two prime benefits of accident investigations are:

- Awareness—Individuals should be made aware of the most frequent types of accidents and the causes of these accidents.
- Factual knowledge—Misconceptions about safety and accidents exist because all the facts may not be known or presented. Unfortunately, these misconceptions and myths continue to persist in spite of refutable data and educational efforts (see Table 3-1).

It is important to have a general understanding of how and why accidents occur, how people are affected by accidents, and how to avoid them. This information should come from

Table 3-1 Safety Myths?

Safety Myths or Not
Lightning never strikes twice in the same place.
A drowning person always comes up for air three times.
If your boat overturns, you should swim to shore.
The first step in saving a drowning person is to swim to them.
It is impossible to stay afloat in water for long with clothes on.
Red is the hunter's best clothing color.
Applying a tourniquet is the best way to stop bleeding.
Numerous cups of coffee will sober up a drunk.
The primary danger from leaking gas is asphyxiation.
Rub snow on frostbite to make circulation return.

© Cengage Learning 2013

data and facts, not suppositions and bias. By the way, all of the statements in Table 3-1 are false.

Our attitudes and values determine the meanings we find in what we observe. A study of some widespread misleading attitudes toward accidents may help show why people react to facts so differently and seem to always have accident problems. Listed below are some misleading beliefs that determine individual attitudes and values about accidents.

- *It won't happen to me.* It is assumed that accidents always happen to other people but won't happen to you.
- *My number's up.* This concept assumes that when your number is up you will get hurt and there is nothing you can do about it. Accidents are determined purely by fate.
- *Law of averages.* Accidents and injuries are due to inevitable statistical laws. Sooner or later everyone gets hurt.
- *Macho concept.* Living dangerously is manly and safety measures are regarded as wimpy.

People react to facts differently. Sometimes education helps to change their beliefs and values. Sometimes civil penalties (fines) modify their behavior. Sometimes nothing changes their beliefs or behaviors.

Accidents Prior to Government Regulation

When the Railway Safety Act was being considered in 1893, a railroad executive said it would cost less to bury a man killed in an accident than to put air brakes on a car. He considered the safety of workers only in monetary terms. He also believed workers assumed liability when they hired on and it was their responsibility to be safe even in unsafe situations. There was a long, hard struggle to provide safeguards to eliminate or reduce accidents and the injuries and damages that result. The struggle was influenced by two mutually opposing considerations: (1) the costs of accident prevention, and (2) moral regard for human life and well being.

A moral consideration for the lives of workers developed because of the number of accidental deaths and injuries. Gradually compromises came about between the benefits and the costs of accident prevention. Many of the larger companies found the mutual consideration and compromise beneficial. The result was fewer walkouts and strikes and more efficient operations. A safe worker wastes less time avoiding hazards. Another benefit to companies has been a reduction in costs of litigation, insurance premiums, and audits by regulatory agencies. Beginning in 1908, when U.S. Steel began its first formal corporate safety program, companies have found that safety programs reduced the costs of doing business. It was common sense. The prime consideration of almost every worker is his or her health and safety. If these are safeguarded in daily activities, the worker is better motivated and more productive.

THE COST OF ACCIDENTS AT WORK

The Division of Vital Statistics reports that accidents are the leading causes of death for persons in their teens and up to age 45. In industry, there is no intent to kill or injure workers yet accidents kill and maim people. Many of the injuries reported as sprains and strains often involved the back. The incidence of fatalities and injuries (along with potential monetary losses) may increase as operations become more complex. The cost of accidents in the workplaces of the United States is approximately $150 billion annually. Some examples of costly accidents are:

- Arco Chemical Company paid $3.48 million in fines as a result of failing to protect workers from an explosion at its petrochemical plant in Channelview, Texas.
- USX's steel-making division paid a $3.25 million fine to settle numerous health and safety violations.
- BASF Corporation agreed to pay a fine of $1.06 million to settle OSHA citations associated with an explosion at a chemical plant that resulted in two deaths and 17 injuries.

Besides the fines, management also incurs costs for safety corrections, medical treatment, survivor benefits, death and burial costs, site insurance, and many other indirect costs. Fortunately, the trend in accident rates in the United States is downward due to the success of the safety movement. In 2001, according to the National Safety Council (NSC), disabling injuries on the job in the United States cost all parties involved, from companies to employees to the government, about $132 billion or about $29,000 per incident. The effect of accidents on profitability is significant. Catastrophic accidents frequently result in damage to the production facility that requires extended downtime for repairs and new construction. Besides this monetary cost, there is the cost of employee morale and its effect upon productivity when a coworker is killed or severely injured. If low morale can devastate professional sports teams, it can also severely affect workplace teams. Few things are as injurious to employee morale as seeing or knowing of a fellow worker injured. Besides worrying about an injured friend or team member, coworkers can't help but think, "That could have been me."

Personal Injury Costs

Multimillion-dollar personal injury awards are not unusual. To the cost of the loss awarded by a court must be added that of the defense. Not every plaintiff in an accident suit has a successful case, but the percentage is substantial. The vulnerability of industry to high costs from accident-related court awards was demonstrated in 1997 in a product-design liability case. A plaintiff family was awarded $262 million from a major automobile manufacturer. A 6-year-old was thrown from a car involved in a traffic accident. The claim alleged the manufacturer of the car was negligent in providing a latch design on a van's rear door that flew open during the accident. (CTDNews, "News Briefs: Workers Win $10.6 Million in Lawsuit," LRP Publications, Volume 6, No. 10, October 1997.) A 27-year-old man working a turn-around in Los Angeles was awarded $4,735,996 to compensate for a severe injury suffered by the man when he was struck by a 630-pound pipe. A 69-year-old man, who lost his right arm when removing scrap from a punch press, was awarded $1,750,000. His lawyer claimed the manufacturer had failed to equip the press with a proper safety device.

Lost Time Costs

According to the NSC, the economic impact of fatal and nonfatal unintentional injuries amounted to $693.5 billion in 2009. This is equivalent to about $2,300 per capita, or about $5,900 per household. These are costs that every individual and household pays whether directly out of pocket, through higher prices for goods and services, or through higher taxes. Approximately 35,000,000 hours are lost in a typical year as a result of accidents. (National Safety Council, Injury and Death Statistics, Injury Facts, 2009)

This is actual time lost from disabling injuries. Often there is spillover from accidents that occurred in previous years that cause lost time in the current year. As an example, a worker exposed to ammonia fumes that suffered permanent damage to his lungs can recover and return to work, but in the following years he may be more susceptible to pneumonia and be unable to work due to illnesses directly related to the previous injury.

Accidental Deaths on the Job

Deaths on the job have decreased over the years, however, they still occur. In a typical year there are 10,400 work deaths in the United States. The deaths are due to a variety of reasons, such as falls, electrocution, drowning, fires, explosions, poison, etc. Table 3-2 lists accidental work deaths by cause for a typical year.

WORK INJURIES

Over exertion, the result of employees working beyond their physical limits, is the leading cause of work injuries. This does not imply that management created brutal working conditions and long work hours. Injury by over exertion is often associated with the temperature of the work environment and the worker's physical condition. Someone in excellent health and physical condition working in 99°F temperatures for a 12-hour shift may over exert them self. Someone 70 pounds overweight doing lifting and climbing is a prime candidate for over exertion. NSC data determined almost 31 percent of all work injuries are caused by over exertion.

Impact accidents involve a worker being struck by or striking against an object. Impact accidents are more frequent during turnarounds when a lot of equipment is being moved, lifted, and transferred around. The next most prominent cause of work injuries is falls. Operators, instrument technicians, and analyzer technicians do a lot of climbing on towers

Table 3-2 Percent of work deaths by cause

Percent of Work Deaths by Cause	
Type of Accidents	**Percent**
Motor vehicle related	37.2
Falls	12.5
Electric current	3.7
Drowning	3.2
Fire related	3.1
Air transport related	3.0
Poison (solid, liquid)	2.7
Water transport related	1.6
Poison (gas or vapor)	1.4
Other	31.6

© Cengage Learning 2013

Source: Bureau of Labor Statistics

and vessels. In the mornings the ladders are wet with dew. They can miss their footing on a damp metal ladder and fall just two rungs to the deck and sprain an ankle. The remaining accidents are distributed fairly equally among the other causes just listed.

Injuries to Body Parts

In order to develop and maintain an effective safety and health program, it is necessary to know not only the most common causes of death and injury but also the parts of the body most frequently injured. The most frequent injuries to specific parts of the body are listed below in order of most frequent injury to least frequent.

- Back *Most common*
- Legs and fingers
- Arms and multiple parts of the body
- Trunk
- Hands
- Eyes, head, and feet
- Neck, toes, and body systems

The back is the most frequently injured part of the body. Most safety and health programs strongly emphasize instruction on how to lift without hurting the back. Back injuries are often due to poor lifting techniques. Being out of shape makes it easier to incur a back injury. Once the back is injured, it becomes more prone to being injured again.

Working in the process industry also makes injuries to legs and fingers more prevalent. The network of pipes and valves and ladders and sharp steel or corroded iron equipment creates a prime habitat for leg and finger injuries. Knees get banged up, legs twisted or sprained, and fingers pinched, cut, abraded, or crushed. These things don't have to happen.

To an alert and careful operator, they won't happen. But the opportunity for them to happen abounds, especially at night and in bad weather.

ESTIMATING THE COST OF ACCIDENTS

Accidents are expensive. To successfully incorporate prevention in the workplace, management must be shown that accidents are more expensive than prevention. To do this, they must be able to estimate the cost of accidents. The costs associated with workplace accidents, injuries, and incidents fall into the broad categories listed in Figure 3-1.

Calculating the direct cost associated with lost work hours involves compiling the total number of lost hours for the period in question and multiplying the hours times the applicable loaded labor rate. The loaded labor rate is the employee's hourly rate plus benefits. Benefits vary from company to company, but typically inflate the hourly wage by 20 to 35 percent. An example of how to compute the cost of lost hours is shown here.

Employee Hours Lost (4th quarter) x Average Loaded Labor Rate = Cost
400 x $30.00 = $12,000

In this example, the company lost 400 hours of work due to accidents on the job in the fourth quarter of its fiscal year. The three employees who actually missed time at work formed a pool of people with an average loaded labor rate of $30 per hour ($25 average hourly wage plus 20 percent for benefits). The average loaded labor rate multiplied times the 400 lost hours reveals an unproductive cost of $12,000 to this company.

By studying company records, management can also determine medical costs, insurance premiums, property damage, and fire losses for the time period in question. All of these costs, taken together, result in a subtotal cost. This figure is then increased by a standard

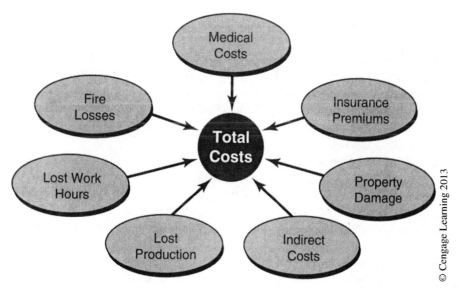

Figure 3-1   Costs Associated with Workplace Injuries

percentage to cover indirect costs to determine the total cost of accidents for a specific time period. The percentage used to calculate indirect costs can vary from company to company, but 20 percent is a widely used.

Following this paragraph the reader will find a partial list of some major industrial accidents. The list has been included to make the point that the potential for accidents is enormous. Manufacturing facilities next to neighborhoods, pipelines running through communities, hazardous cargo vehicles, and vessels on roads, rails, and waterways throughout the country—all of these situations are opportunities for an accident. What will prevent those accidents from occurring are knowledgeable and alert workers. Accidents don't have to happen.

PARTIAL LIST OF MAJOR INDUSTRIAL ACCIDENTS (1917 TO 1985)

(Source: *Marrin Abraham, The Lessons of Bhopal: A Community Action Resource Manual on Hazardous Technologies,* IOCO, Penang, Malaysia, September 1985.)

December 6, 1917, Halifax, Nova Scotia, Canada:

A French ship carrying about 1,000 tons of ammunition collided with a Belgian steamship, setting off explosions that destroyed a two-square-mile area of Halifax and damaged nearby piers. Some 1,654 people were killed.

September 21, 1921, Oppau, Germany:

The biggest chemical explosion in German history occurred at a nitrate manufacturing plant about 50 miles south of Frankfurt. The blast destroyed the plant, a warehouse, and leveled houses four miles away in the nearby village of Oppau. At least 561 lives were lost and some 1,500 people injured.

October 20, 1944, Cleveland, Ohio:

A poorly designed liquefied natural gas tank belonging to the East Ohio Gas Company developed structural weakness, resulting in a massive explosion. The ensuing blast and fire claimed some 131 lives.

April 16, 1947, Texas City, Texas:

A freighter, the "Grand Camp," carrying 1,400 tons of ammonium nitrate fertilizer exploded after fire broke out on board. The initial explosion set off a series of secondary explosions that destroyed much of Texas City. The blast rattled windows 150 miles away and the leaping flames also destroyed a nearby Monsanto plant producing a styrene. The next day another freighter, the "High Flyer," also loaded with nitrates, exploded in the same harbor. Some 576 people were killed and 2,000 others seriously injured.

June 1, 1974, Flixborough, UK:

A large twenty-inch bypass pipe carrying caprolactum at a NYPRO Ltd. plant leaked, resulting in the escape of a large amount of the chemical. The caprolactum cloud exploded, setting off a fire over 20 acres of land. The blast killed 28 people, injured 36, and 3,000 others were evacuated; the blast also leveled every building on the 60-acre plant site.

July 10, 1976, Seveso, Italy:

An uncontrolled exothermic reaction in a reactor at the Hoffman-La Roche Givaudan chemical plant caused a major explosion. The ensuing release of only 10–22 pounds of toxic tetrachlorodibenso-p-dioxin contaminated soil and vegetation over 4,450 acres of land, and killed over 100,000 grazing animals. Although there were no immediate injuries or loss of human lives, over 1,000 residents were forced to flee, and many children subsequently developed a disfiguring rash called chloracne.

November 10, 1979, Mississauga, Ontario, Canada:
A total of 21 railroad cars carrying caustic soda, chlorine, propane, styrene, and toluene derailed. Three of the railroad cars carrying propane and toluene exploded and caught fire while a fourth railroad car carrying chlorine ruptured and its contents also caught fire. No lives were lost, but eight fire fighters were injured and 250,000 people evacuated.

December 2–3, 1984, Bhopal, India:
The escape of some 40 tons of MIC (methyl isocyanate) gas from a Union Carbide pesticide production plant in the Indian city of Bhopal led to the world's worst industrial disaster. At least 2,500 people were killed, 10,000 seriously injured, 20,000 partially disabled, and 180,000 others adversely affected in one way or another. Some 150,000 people are reported to be still suffering from the adverse effects of the Bhopal catastrophe.

THE HUMAN FACTOR

Most industrial accidents that happen around the world are caused by human error. Human error is as much an indictment of company organization and management as of the employees involved. The twentieth century's worst industrial disasters—Bhopal, Three Mile Island, and Chernobyl—helped clarify the complex chain of systemic problems that lead to human error. The accidents provided numerous checklists and case studies for control room and equipment design. Almost every mishap can be traced ultimately to human error, either on the part of the person immediately involved in a mishap, a designer who made a mistake in a calculation, a worker incorrectly manufacturing a product, inadequate management or procedures, etc. Poor design of machinery, equipment, and control systems may be a significant cause of accidents by contributing to worker error. On the other hand, human error may cause well-designed equipment to fail. But what do we mean by the **human factor** that leads to human error? There are numerous reasons why errors occur. Some factors contributing to human error are:

- Skill level
- Technician fatigue
- Failure to follow procedures
- Inadequate procedures
- Lack of clear communication protocols
- Poorly designed equipment
- Improper understanding of ergonomic issues by management

HUMAN ERROR

Human error can be defined as *an action that is inconsistent with established behavioral patterns (speeding ticket, public intoxication, etc.) considered normal or that differs from prescribed procedures.* Errors can be divided into two categories: **predictable** and **random**. Predictable errors are those which experience has shown will occur again if the same conditions exist. Predictable errors can be foreseen because their occurrence has taken place more

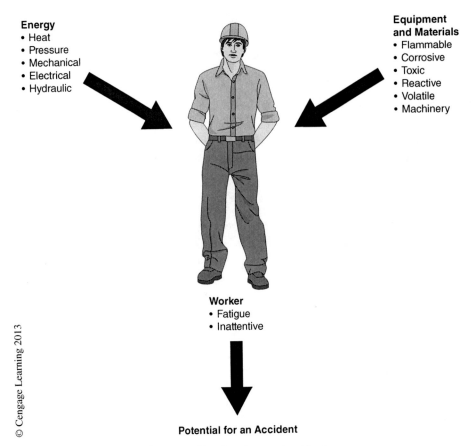

Energy
● Heat
● Pressure
● Mechanical
● Electrical
● Hydraulic

**Equipment
and Materials**
● Flammable
● Corrosive
● Toxic
● Reactive
● Volatile
● Machinery

Worker
● Fatigue
● Inattentive

Potential for an Accident

© Cengage Learning 2013

Figure 3-2 Interaction of Factors That Lead to an Accident

than once. For example, people will generally tend to follow procedures that involve minimal physical and mental effort, discomfort, or time. Any procedure contravening these basic principles is certain to be illegally modified or ignored at some time by the persons carrying it out. Many workers take short cuts or often ignore work rules they consider a nuisance. Human nature makes this kind of error predictable. Some errors are caused by the interaction of two or more factors (see Figure 3-2) and often cannot be predicted. Random errors are unpredictable and cannot be attributed to a specific cause. For example, a highly competent operator may be annoyed by a fly and swat at it. In doing so, he bumps a critical control device that causes a unit upset. There are fewer types of random errors than of predictable errors.

In the workplace errors are further sorted into two types, **errors of omission** and **errors of commission.**

● An error of omission is the failure to perform a required function. A step is left out of a prescribed procedure, intentionally or accidentally, or a sequence of operations may not be completed. In some instances, intentional omissions may be due to procedures that are too long, badly written, contrary to normal tendencies and actions, or not readily understood.
● An error of commission is performing a function not required, such as unnecessarily repeating a procedural step, adding unnecessary steps to a sequence, or doing an erroneous step.

36

Some following paragraphs discuss the various types of human errors.

Designing and Planning Errors

The person who designs equipment or plans an operation can make errors such as (1) miscalculations, (2) failing to remove or control a hazard, and (3) failing to incorporate safeguards to prevent accidents or protect personnel. Some of these errors we hear about frequently, especially those concerning automobile recalls for potential safety defects. When a designer or planner cannot completely eliminate a hazard or the possibility of an accident, they must attempt to minimize the possibility that operators will commit errors leading to mishaps. In effect, the designer, through foresight, must attempt to make the system "idiot-proof," although we all know about the inevitability of Murphy's Law.

Designers' errors may invite mistakes in reading dials or gauges or may lengthen operator reaction times. With a large array of widely spaced instruments or several consoles, reviewing them often might require an operator to have two heads. A design error can also be one that violates a normal tendency or expectancy. Employees expect that on a vertically numbered instrument, the higher-value numbers will be at the top; on a circular dial they expect values to increase clockwise. An improper design can unduly stress the operator. Poor design may require employees to wear burdensome protective respiratory equipment for long lengths of time that result in fatigue that may lead to errors. Other fatigue-producing causes are:

- Glare
- Inadequate lighting
- Uncomfortable chairs
- Vibration or noise
- Unusual positions in which to operate
- Closeness to hot surfaces

All of these poor design conditions can produce fatigue, stress and a lack of motivation.

Procedural Errors

Procedural errors can turn a minor emergency into a major emergency. At such times, personnel are almost always in a state of stress and extremely susceptible to committing errors. No matter how calm the individual may appear during an emergency their ability to make decisions will be impaired and they will be more susceptible to making errors. Procedural errors can be minimized if the procedures are:

- Clearly written
- Concise
- Have backout steps in case of error or emergency steps
- Warnings and explanations about critical steps

Two-Person Concept

To minimize the possibility of human error in any procedure involving a nuclear device, the U.S. Department of Defense has developed the *two-person concept*. Two or more persons, each capable of undertaking the prescribed task and of detecting an incorrect or unauthorized step in a procedure, are assigned to the task. One person accomplishes the procedural step and the other checks the action to verify it has been done correctly. It is not necessary that both persons have equal knowledge of the task, only that each is able to detect and ensure that the actions of the other are correct according to procedure. Commercial airline pilots use this concept before take-off and landings. Some situations in the processing industry may require the two-person concept.

SAFETY PROMOTION TO PREVENT ERRORS

In former years, many efforts to promote safe practices consisted almost entirely of a campaign that alerted employees to the hazards in their workplace and urged them to work safely. The modern concept is that hazards that cannot be eliminated should be controlled, first by design and then by procedural means. Procedural means consist in relying on employees to perform tasks properly and safely. Use of procedures is a less desirable means of accident prevention than is good design but because all hazards cannot be eliminated by design companies rely on the safe practices of their workers. To ensure safe work practices by their employees, companies must have an effective safety promotion plan.

"Why won't they work safe? Why do our workers take shortcuts?" These are just a few questions asked by management from the first-line supervisor all the way to top executives in an organization. In turn, the employee asks, "Why doesn't management support safety? All I hear from management is get it done! Management says nothing is more important than safety, but it seems like lip service to me." Both groups have valid points. When looking at an organization's safety culture one must look at the commitment of every employee and their motives for having a safe work environment. Safety must be the responsibility of the individual. The individual is made up of top executives down through the organization to the newly hired process technician.

A very strong interest in behavior-based safety emerged at the close of the 90s decade. It has often been said that every worker is a manager. Workers can be expected to manage their situation to their personal advantage. If they can get away with shortcuts that save time and effort but increase risk, some workers take the shortcuts. In behavior-based safety, the culture, not the Occupational Safety and Health Administration (OSHA), drives the safety process. Behavior-Based Safety (BBS) focuses intervention on observable behavior, directing and motivating managers and workers through *activators* and *consequences*. In BBS, both workers and management must participate actively and buy into the process. This process seeks to produce a safety culture unique to a specific work site. This approach focuses on observable behaviors, provides positive activators to motivate workers (changes work conditions, modifies the task, includes extra manpower, etc.), and applies continuous interest and evaluation. Positive activators are those which provide the workers with a sense of empowerment, freedom, and control. These are longer lasting than negative activators like fear of punishment.

Each worker is personally responsible for their own safety and for observing an unsafe act or unsafe behavior and pointing it out to the persons committing them. In addition, it means allowing fellow workers the freedom to point out things being done that could cause an accident. It is important for both parties, workforce and management, to not create an adversarial culture. A successful behavior-based safety program is not weighted to one side or the other in areas of responsibility. If an employee or someone considered to be at the bottom of the organizational list observes unsafe behavior and feels that they cannot approach the situation because of retaliation, the culture is wrong and the creation of a behavior-based safety program will fail.

Safety Regulations

Safety regulations are as old as the first human communities. Any place where people worked in groups, they developed rules to protect themselves while they worked. Go to any third world country where there is no OSHA, and you will find humans have created safety rules to protect themselves. These rules are codes of conduct to avoid injury and damage. Employers have been cited and fined under the OSHA Act where employees were injured or killed because of failures to enforce safe work rules. Safety rules may have been published prohibiting horseplay yet some supervisors have routinely witnessed horseplay and failed to stop it. In such a case, when a worker is injured due to horseplay, the company may be liable to a personal injury suit. The injured worker would be entitled to Workers' Compensation, plus one or more people may lose their jobs.

Employee Participation in Safety

After the OSHA standards went into effect in 1971, OSHA inspectors visited many plants after their unions or individuals charged that numerous imminent dangers existed. The workers in any process are the ones most aware of the hazards in their process and they want the hazards mitigated or eliminated because their lives and health are at stake. Workers are excellent sources of information about hazards.

Employees can participate in any safety effort in numerous ways (see Figure 3-3). One way is to report equipment design or performance deficiencies. Many workers fail to report

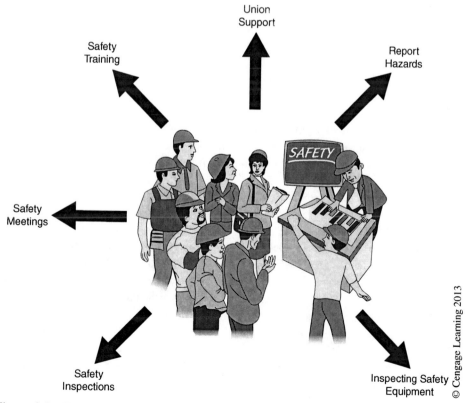

Figure 3-3   Employee Participation in Safety

such deficiencies to their supervisors for correction for several reasons. Often it is because they believe the presence of such hazards is normal and they accept the potentially injurious condition. Some workers simply don't recognize the danger. Others may recognize it but will do nothing unless they are rewarded in some way.

Another way an employee can participate in safety is through their union. Management and unions frequently disagree on production goals, work rules, and numerous other things, but they do share the common concern of safety. Accident prevention is the foremost area where management and unions cooperate very well together. Historically, a major reason for the birth of industrial unions has been the health and safety of its members. Management will often find the union representative to be a strong supporter of management safety efforts.

Safety training is another way employees participate in safety. Through training they learn how to identify the different types of hazards. The training should begin with the new employee and continue until the employee retires. The type of training, frequency, and material presented will vary with employment. Safety training should be given to all new employees regardless of previous experience. Topics in this training should include information on and review of:

- The company safety rules and practices
- The employee duties and rights under the OSHA Act or state safety codes
- Emergency signals and their meanings
- How to use PPE and emergency equipment
- How to summon help in times of need

Their immediate supervisor or a designated experienced and knowledgeable coworker should indoctrinate the new employee further. This indoctrination should include safety topics listed here:

- Hazards in the operations in which the employee will participate
- Safeguards that have been provided and precautionary measures to be taken against those hazards
- Locations of emergency exits, fire extinguishers, first aid kits, and other emergency equipment workers might use
- Procedure to follow in the event a specific type of emergency occurs
- How to report hazards or defective equipment
- The need for good housekeeping

Safety (Tailgate) Meetings

Even experienced employees can benefit from regular on-the-job safety training. One of the most effective means is a crew's daily safety meetings, often called *tailgate meetings*. Members of work teams (operations crews, instrumentation and analyzer crews, maintenance crews, etc.) attend short 15 minute safety meetings each morning before beginning their work and the subject matter is pertinent to their activities. The meeting can be lead by the crew supervisor or may be rotated among crew members. Informal meetings should allow attendees to contribute comments. Such meetings might include:

- Instructing employees on new types of equipment, their uses, capabilities, hazards, and safeguards
- Listing the precautionary measures required for a task they will undertake today

- Discussing accidents or unsafe practices recently witnessed
- Reporting of unsafe equipment, conditions, or practices

SUMMARY

Accidents are the leading causes of death for persons in their teens and up to age 45. The cost of accidents in the workplaces of the United States is approximately $150 billion annually. Work accidents are expensive. Besides the fines, management also incurs costs for safety corrections, medical treatment, survivor benefits, death and burial costs, plus many other indirect costs. The value of lost production due to accidents can exceed those listed above.

Most industrial accidents are caused by human error. However, the error is often due as much to a company's organization and management as to its workers. It is important to have a general understanding of how and why accidents occur, how people are affected by accidents, and how they can be prevented. This understanding should come from data and facts, not suppositions and bias. There was a long, hard struggle to provide safeguards to eliminate or reduce accidents and the injuries and damages that result. The struggle was influenced by two mutually opposing considerations: (1) the costs of accident prevention, and (2) moral regard for human life and well being.

Errors can be divided into two categories: *predictable* and *random*. Predictable errors are those which experience has shown will occur again if the same conditions exist. Random errors are unpredictable and cannot be attributed to a specific cause. Many random errors can be included under a general safeguard, whereas for a predictable error a specific safeguard may be provided.

Because all hazards cannot be eliminated by design, companies must rely on the safe work practices of their employees. To ensure safe work practices of their employees, companies will have to have an effective safety promotion plan. The plan should include employee participation in safety and safety training.

REVIEW QUESTIONS

1. Two benefits of accident investigations are _____ and _____.

2. Explain why the injuries keep happening from the same incidents at the same plant.

3. List the five leading causes of accidental death in the United States.

4. List four elements that contribute to the overall cost of an accident.

5. In a typical year there are _____ work deaths in the United States.

6. Most worldwide industrial accidents are caused by _____.

7. The three most frequent injuries to the body are to the _____, _____, and _____.

8. (T/F) Errors and accidents frequently occur because equipment is designed for a statistical (average) worker.

9. _____ is an action that is inconsistent with the established behavioral patterns considered normal or that differs from prescribed procedures.

10. Predictable errors are _____:
    a. Those that leave out a step in a procedure
    b. Those that add an extra step to a procedure
    c. Those which experience has shown will occur again if the same conditions exist
    d. Those that cannot be attributed to a specific cause

11. _____ are non-predictable and cannot be attributed to a specific cause.
    a. omission
    b. predictable
    c. commission
    d. random

12. An error of _____ is the failure to perform a required function.

13. An error of _____ is performing a function not required.

14. A _____ is one that violates the normal tendency or expectation, such as on a circular dial the values decrease clockwise.

15. (T/F) Safety must be the responsibility of the individual, not management.

16. List three ways the employee can participate in safety.

17. Tailgate meetings usually take about _____ minutes and can be conducted by a supervisor or crew member.

EXERCISES

1. COST OF AN ACCIDENT

Instructions

After reading the paragraph below, answer the questions at the end.

An outside operator was working with 90-pound steam (331°F) using cotton gloves instead of the required heat resistant-gloves. The operator received a third-degree burn to his left hand, was rushed to medical, and then taken by plant ambulance to a hospital. After receiving treatment, the operator was sent home. As an outside operator, his job required the use of two hands to perform his duties. He did not return to work for three weeks. The operator made $31 per hour and worked 40 hours a week. Overtime to replace him cost $46.50 an hour. Three people (one engineer, one safety officer, and one operator) conducted an accident investigation that took two hours at an average cost of $43 per hour each. The completed report was read and discussed at Monday morning's management meeting taking up 30 minutes of 5 manger's time at a cost of $55 per hour. The hospital bill for emergency treatment cost the plant $670. Two operators, who were also EMTs, left their jobs to drive the ambulance to the hospital and remained at the hospital two hours each before returning back to the plant. They make $31 per hour. The accident report will have to be included in monthly injury statistics and annual injury statistics plus entered in the OSHA 3000 log. The cost for this was $28 total. The steam burn occurred in November and pushed the plant lost time injury total high enough for the year to warrant an OSHA site inspection. Two OSHA inspectors came out and met with the plant manager, operations manager, safety and health supervisor, unit engineer, unit lead operator, and three operators of the unit where the operator received the steam burn. OSHA was in the plant for three hours for a total cost of $876.

   a. Calculate the total cost to this production site for this steam burn that took less than 5 seconds to occur.

   b. What is the one thing the operator could have done to have prevented this accident?

2. Go on the Internet and research one of the accidents described in the chapter. Find the cost of that accident and write a one page report detailing the cost.

3. Write a one page report justifying whether you support or disagree with the following scenario. List five or more reasons why you want the worker to be disciplined or not to be disciplined. **Scenario**: A worker who has been in the plant and on his unit for 18 years made an error of omission, damaged a $50,000 pump, and caused a line break that released 15 gallons of chemical onto the unit pad. The worker had only one other accident ten years ago, gets along with everyone, and has always been considered very reliable. Management wants him suspended from work with no pay for 30 days. The union disagrees because of his good work record. You be the judge and pass down a verdict.

RESOURCES

www.wikipedia.org

www.osha.gov

www.YouTube.com

CHAPTER 4

Hazard Recognition

Learning Objectives

At the end of this chapter, the student should be able to:

- *List five physical hazards.*

- *Explain the difference between a combustible and flammable liquid.*

- *Explain what is meant by a chemical's* lower explosive limit *and* upper explosive limit.

- *Explain why water reactive chemicals are considered a physical hazard.*

- *Write definitions for* health hazard *and* physical hazard.

- *Describe six health hazards.*

- *Explain the following terms:* corrosive, sensitizer, mutagen, carcinogen, *and* teratogen.

INTRODUCTION

Ask anyone working in processing and manufacturing industries what is the first step in accident prevention, and most will respond that it is knowing the hazards associated with their tasks. A typical plant will have numerous hazards. There are chemical hazards posed by the processing, storing, and handling of chemicals. There are physical hazards posed by the physical properties of chemicals. Then, there are ergonomic, biological, and security hazards. Ergonomic hazards, such as carpal tunnel syndrome and back injuries, will not be discussed except to say that back injuries from improper lifting techniques are the greatest ergonomic hazard to operations personnel. Biological hazards (Legionnaire's disease, mosquito-borne diseases, etc.) generally are not very common in the process industry. Security hazards are discussed in the last chapter of this book. Chemical and physical hazards are the predominate hazards to process employees and are the focal points of this chapter.

Safety hazards related to the physical characteristics of a chemical can be defined in terms of testing requirements, such as ignitability; however, health hazard definitions are less precise and more subjective. There have been many attempts to categorize effects and to define them in various ways. Generally, the terms **acute** and **chronic** are used to delineate between effects on the basis of severity or duration. Acute effects occur rapidly as a result of short-term exposures. Chronic effects occur as a result of long-term exposures.

Acute effects referred to most frequently are those defined by the American National Standards Institute (ANSI) and are irritation, corrosivity, sensitization, and lethal dose. Similarly, the term chronic effect is often used to cover only carcinogenicity, teratogenicity, and mutagenicity. Acute exposures to airborne chemicals are typical of transportation accidents, fires, or accidental releases at chemical manufacturing or storage facilities. Acute airborne exposures normally occur when there is an accidental release and personnel are unable to evacuate quickly enough. Chronic exposures usually are associated with normal plant operations when personal protective measures are not adequate, such as when the concentration exceeds the permissible exposure limit (PEL). Chronic exposure may be due to a careless attitude by the employee or exposure to chemical from undetected slow leaks of gases and volatile chemicals that blend in with the background odor of the unit. There is no list of hazardous materials. Instead, the Occupational Safety and Health Administration (OSHA) has defined two categories of hazardous materials, and they are **health hazards** and **physical hazards**. If a material meets one of OSHA's definitions, it is considered to be a hazardous chemical.

PHYSICAL HAZARDS

Physical hazards include noise, vibration, extremes of temperature, compressed gases, combustible and flammable chemicals, pyrophorics, explosives, oxidizers, and reactive materials. Process employees are exposed to physical hazards on a daily basis because they work outside on the unit among pipes containing compressed gases, fluids under high temperatures, and flammable and explosive chemicals. Examples of some these hazards on a process unit might include:

- High pressure and temperature steam lines at 600 psig and 488.8°F
- 16 inch diameter lines full of gasoline
- Plant air lines at a pressure of 120 psig
- Natural gas lines for furnace burners
- Compressed gases at 2400 psig used for analyzers

Any chemical that has a physical hazard as defined in the following pages constitutes an OSHA defined hazardous chemical. The hazards of noise and vibration are discussed in greater detail in a later chapter.

Combustible Liquids

A combustible liquid is any liquid having a flash point at or above 100°F (37.8°C) but below 200°F (93.3°C). The exception to the preceding sentence is any mixture having components with flash points of 200°F (93.3°C), or higher, the total volume of which make up 99 percent or more of the total volume of the mixture.

Compressed Gases

A compressed gas is a gas or mixture of gases having, in a container, an absolute pressure exceeding 40 psi at 70°F (21.1°C). It can also be a gas or mixture of gases having, in a container, an absolute pressure exceeding 104 psi at 103°F (54.4 °C) regardless of the pressure at 70°F (21.1 °C). Finally, it can be a liquid having a vapor pressure exceeding 40 psi at 100°F (37.8°C) as determined by ASTM D-323-72.

Explosives

An explosive is a chemical that causes a sudden, almost instantaneous release of pressure, gas, and heat when subjected to sudden shock, pressure, or high temperature.

Flammables

There are several types of flammables based on their physical phase. An **aerosol flammable** is an aerosol that, when tested (method described in 16 CFR 1500.45) yields a flame projection exceeding 18 inches at full valve opening, or a flashback (a flame extending back to the valve) at any degree of valve opening. A **gas flammable** is (1) a gas that, at ambient temperature and pressure, forms a flammable mixture with air at a concentration of 13 percent by volume or less; or (2) a gas that, at ambient temperature and pressure, forms a range of flammable mixtures with air wider than 12 percent by volume, regardless of the lower limit. A **liquid flammable** is any liquid having a flash point below 100°F (37.8°C), except any mixture having components with flash points of 100°F (37.8°C) or higher, the total of which makes up 99 percent or more of the total volume of the mixture. And finally, a **solid flammable** is (1) a solid, other than a blasting agent or explosive (defined in CFR 1910.109a) that is liable to cause fire through friction, absorption of moisture, spontaneous chemical change, or retained heat from manufacturing or processing, or (2) which can be ignited readily and when ignited bums so vigorously and persistently as to create a serious hazard. A chemical shall be considered to be a flammable solid if when tested by the method described in 16 CFR 1500.44, it ignites and burns with a self-sustained flame at a rate greater than one-tenth of an inch per second along its major axis.

Flash Point

The flash point is the lowest temperature at which vapors on the surface of a liquid in a test apparatus will ignite and burn when an ignition source is placed in the vapor space. However, at the flash point, the fire would last only long enough to burn the vapors and then go out. Different chemicals have different flash points. For instance, benzene has flash point of 12°F, while gasoline has the flash point of -45°F. How does flash point information warn us about a hazard? Generally, the lower the flash point the greater the flammability hazard posed by that chemical. For safe transportation and storage of chemicals, the Department of Transportation (DOT) and the National Fire Protection Association (NFPA) have classified chemicals into various groups depending on their flash points. DOT regards a chemical as flammable if the flash point is below 100°F and combustible if the flash point is at or above 100°F.

Explosive Limits (Flammability Limits)

Any combustible or flammable material has a lower flammability limit (LFL) and an upper flammability limit (UFL). These are also more commonly known as the lower explosive limit (LEL) and upper explosive limit (UEL). The lower explosive limit is the minimum concentration of the chemical in air at or above which the material will support ignition. Any concentration below this and the mixture is too lean to burn. Similarly, the upper explosive limit is the highest concentration above which the material will not support combustion. Any concentration above this and the mixture is too rich to burn.

Many chemicals have their LEL-UEL range between 1 to 5 percent, but there are several exceptions. For instance, gasoline's LEL-UEL range is 1.5 to 7.6 percent. Acetylene's range is much wider, 2.5–100 percent. Generally, the wider the LEL-UEL range, the higher the fire hazard posed by that chemical. What is the practical use of knowing LEL-UEL? Many plants use LEL meters to check areas before issuing hot work permits for the area. Typically, many plants consider a reading of ten percent or less on the LEL meter as an acceptable limit to issue a permit for work or entry. Some explosive limits are shown in Table 4-1. For a better understanding of explosive limits, study Figure 4-1.

Organic Peroxides

Organic peroxides are an organic compound that contains the bivalent oxygen structure (-O-O-). They release oxygen readily and are strong oxidizing agents. Because they release oxygen readily they are a source of oxygen needed to complete the fire triangle. They are fire hazards when in contact with combustible materials, especially at high temperatures. They are also unstable and may decompose if concentrated.

Oxidizers

Oxidizers are a chemical other than a blasting agent or explosive that initiates or promotes combustion in other materials thereby causing fire either of itself or through the release of oxygen or other gases. These are chemicals that either spontaneously evolve oxygen at room temperature or under slight heating. This term includes chlorates, peroxides, perchlorates, nitrates, etc.

Table 4-1   Explosive Limits

Explosive Limits—Upper and Lower Volume Percent Substance in Air		
Substance	**LEL**	**UEL**
Hydrogen	4%	75%
Gasoline	1%	8%
Hydrogen sulfide	4%	46%
Propane	2%	11%
Butane	2%	9%
Hexane	1%	7%
Ethylene	2%	29%
Propylene	2%	12%

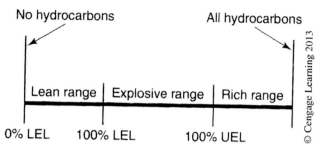

Figure 4-1   Diagram of Explosive Limits

Pyrophorics

A pyrophoric is a chemical that will ignite spontaneously in air at a temperature of 130°F (54.4°C) or below. Pyrophoric materials are chemicals that are capable of igniting in air at ambient temperatures. Earlier, we mentioned alkyls, sodium, potassium, and lithium. These reactive materials can also be considered pyrophoric. Many substances, although otherwise non-reactive, become pyrophoric when exposed to air in a fine powder form. For example, carbon (coke) granules are not pyrophoric, but if the granules are ground to a very fine powder, that same coke becomes pyrophoric. Many deposits formed within process vessels in the absence of oxygen become pyrophoric when exposed to air when the vessel is opened up for inspection or repairs.

Reactive (Unstable) Chemicals

A reactive chemical is a chemical or mixture of chemicals that will vigorously polymerize, decompose, condense, or will become self-reactive under conditions of shocks, pressure, or temperature. Many chemicals such as aluminum alkyls, and alkali metals such as sodium, potassium, and lithium, are highly reactive with water. They must be stored away from a damp or moist environment or an explosion may occur. The nitrogen blanket used for the storage tanks of these chemicals must have a very low dew point; about -40°C. Hoses for loading and unloading such chemicals must be completely dry and be pressure checked before loading chemicals with them. When loading is complete, they must be purged thoroughly.

Specific Gravity

Specific gravity is the comparison of the density of solids and liquids to water and of gases to air. This value reveals how heavy a liquid is in comparison to water. Water has specific gravity of 1.0. If the specific gravity of a liquid is 0.7, then the liquid is lighter than water and (if not miscible) will float on water. The greater the specific gravity, the heavier the chemical and the greater pressure exerted on the containing vessel. The specific gravity of gases is in relation to air, which has a specific gravity of 1.0. Gases heavier than air have a specific gravity greater than 1.0. Most hydrocarbon gases are heavier than air. So, how can knowing the specific gravity of a gas help you with hazard recognition and the proper response to the hazard? If you see a white cloud hugging the ground in an area where there are no steam lines and the cloud is around a butane tank, chances are the cloud is butane vapor visible because it has absorbed some moisture from the air. This vapor cloud is looking for a source of ignition. Butane, being heavier than air, remains near the ground. Knowing the density of a gas compared to air will tell you if that gas, being

Table 4-2   Specific Gravity of Some Common Gases

Gas	Specific Gravity
Air	1.0
Nitrogen	0.967
Oxygen	1.105
Hydrogen	0.069
Methane (natural gas)	0.554
Propane	1.552
Hydrogen sulfide	1.176

© Cengage Learning 2013

lighter than air, will rise upward and dissipate, or being heavier, tend to hug the ground and accumulate. Table 4-2 shows the specific gravity or density of some common gases found in the processing industry.

Water-Reactive

Chemicals that are water reactive generally react with water to release a gas that is either flammable or presents a health hazard. An example would be the alkali metals lithium or sodium which reacts with water to form hydrogen gas, which is flammable.

Figure 4-2 is a graphic representation of the physical hazards we have discussed.

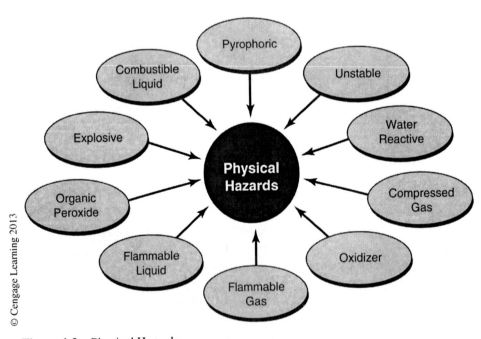

© Cengage Learning 2013

Figure 4-2   Physical Hazards

HEALTH HAZARDS

Health hazards are defined as those chemicals for which there is scientific evidence demonstrating that acute or chronic health effects may occur if employees are exposed to that chemical. A significant amount of the hazards in the processing industry are of the chemical type. For chemical health hazards, the Material Safety Data Sheet (MSDS) is the process employee's best friend for information about a chemical. Like physical hazards, there are a variety of health hazards, some of which have been mentioned in the previous chapter.

Assume you work on a process unit that uses monomethaneolamine and benzene in its process. Just those two chemicals expose you to the following health hazards:

- Carcinogen (cancer causing agent)
- Sensitizer
- Allergen
- Corrosive
- Irritant
- Neurotoxin
- Teratogen
- Mutagen

Carcinogen (cancer causing agent) — A chemical is considered to be a carcinogen if:

- It has been evaluated by the International Agency for Research on Cancer (IARC) and found to be a carcinogen or a potential carcinogen.
- It is listed as a carcinogen or potential carcinogen in the Annual Report on Carcinogens published by the National Toxicology Program's (NTP) latest edition.
- It is regulated by OSHA as a carcinogen.

Corrosives — Corrosives are chemicals that cause visible destruction of or irreversible alterations in, living tissue by chemical action at the site of contact. The pH of a chemical is a good indicator of its corrosivity. The pH scale spans from 0 to 14, with 7 being neutral. A value above 7 is basic, a value below 7 is acidic.

Highly Toxic — A chemical is highly toxic if it falls within any of the following categories: (1) A chemical that has a median lethal dose (LD_{50}) of 50 milligrams or less per kilogram of body weight when administered orally to albino rats weighing between 200 and 300 grams each; (2) A chemical that has a median lethal dose (LD_{50}) of 200 milligrams or less per kilogram of body weight when administered by continuous contact for 24 hours (or less if death occurs within 24 hours) with the bare skin of albino rabbits weighing between 2 and 3 kilograms each; (3) A chemical that has a median lethal dose (LD_{50}) in air of 200 parts per million by volume or less of gas or vapor, or 2 milligrams per liter or less of mist, fume, or dust, when administered by continuous inhalation for one hour (or less if death occurs within one hour) to albino rats weighing between 200 and 300 grams each.

Irritant — An irritant is a chemical which is not corrosive, but which causes a reversible inflammatory effect on living tissue by chemical action at the site of contact. A chemical *—goes away* is a skin irritant, if when tested on the intact skin of albino rabbits by testing (method 16 CFR ISOO.4l) for four hours exposure or by other appropriate techniques, it results in an empirical score of five or more. A chemical is an eye irritant if so determined by test (16 CFR ISOO.42) or other appropriate techniques.

Sensitizer — A sensitizer is a chemical that causes a substantial proportion of exposed people or animals to develop an allergic reaction in normal tissue after repeated exposure to the chemical.

Neurotoxin — A neurotoxin is a poison that acts on the nervous system. Two common ones in petrochemical plants are phenol and tert-butylcatechol.

Mutagen — A mutagen is a chemical that has or is suspected to have the properties that can change or alter the genetic structure of a living cell. Mutagens can make a person more susceptible to cancer.

Teratogen — A teratogen is a substance that has an adverse effect on a human fetus. It can cause the fetus to have deformities from the normal. A good example is a two-headed calf.

SUMMARY

A typical manufacturing plant may have numerous hazards. There are chemical hazards or the hazards posed by the storing, processing, and handling of chemicals. There are also physical hazards posed by the physical properties of chemicals, such as a compressed gases, explosives, and oxidizers. Generally, the terms *acute* and *chronic* are used to delineate between effects on the basis of severity or duration. Acute effects occur rapidly as a result of short-term exposures. Chronic effects occur as a result of long-term exposures.

Physical hazards include noise, vibration, extremes of temperature, compressed gases, combustible and flammable chemicals, explosives, oxidizers, and reactive materials. Any chemical that has a physical hazard, as defined in the follow paragraph, constitutes an OSHA defined hazardous chemical.

Health hazards are defined as those chemicals for which there is scientific evidence demonstrating that acute or chronic health effects may occur if employees are exposed to that chemical. A significant amount of the hazards in the processing industry are of the chemical type. For chemical health hazards, the MSDS yields the best information about the chemical.

REVIEW QUESTIONS

1. Two biological hazards are _____ and _____.

2. _____ effects occur rapidly due to short-term exposures.

3. _____ effects occur over a period of time as result of long-term exposures.

4. (T/F) OSHA has a list of all hazardous materials.

5. List five physical hazards.

6. As a rule, _____ liquids have a flash point at or above 100°F.

7. _____ have a flash point below 100°F.

8. Explain how a chemical's flash point warns a worker about its hazard.

9. Explain what is meant by a chemical's *lower explosive limit* and *upper explosive limit.*

10. (T/F) Generally, the wider the LEL-UEL range, the higher the fire hazard posed.

11. _____ promote combustion in other materials and can cause a fire.

12. Explain how specific gravity alerts an operator to the hazardous nature of a chemical.

13. _____ are those chemicals for which there is scientific evidence demonstrating acute or chronic health effects.

14. List six health hazards.

15. A _____ is a cancer-causing agent.

16. _____ are chemicals that cause visible destruction or damage to living tissue.

17. A _____ is a poison that acts on the nervous system.

18. A _____ is a substance that has a damaging effect on the human fetus.

EXERCISES

1. The specific gravity of air is 1.0. Gases with a specific gravity less than one are lighter than air; those with a specific gravity greater than one are heavier. Refer to Table 4-2 for the specific gravity of the following gases: propane, hydrogen, butane, and hydrogen sulfide. Now, assume you are an operator on a unit that has all of those gases, and that one week a leak in a flange or valve released a large amount of propane. Answer these two questions: (1) based on propane's specific gravity, will the released gas disperse rapidly upward into the air, and (2) is the gas flammable. Answer the same questions for hydrogen, propane, and hydrogen sulfide.

2. Gasoline is a blend of several hydrocarbon chemicals plus a few additives. Research on the Internet the compounds that make up gasoline and then find out if those compounds have any health hazards, such as carcinogenic, mutagenic, etc. Write a one page report.

RESOURCES

www.wikipedia.org

www.osha.gov

www.YouTube.com

www.cdc.gov

www.cdc.gov/niosh/

www.toolboxtopics.com

CHAPTER 5

Toxic Hazards and Blood-Borne Pathogens

Learning Objectives

Upon completion of this chapter, the student should be able to:

- *Write a definition of a toxic substance.*

- *List the seven mechanisms by which toxic substances can cause injury.*

- *Explain how latency is problem in identifying toxic chemicals.*

- *Explain how corrosives affect the body.*

- *Explain how systemic poisons damage the body.*

- *Explain the four routes of entry of toxins into the body.*

- *Explain which route of entry of toxins into the body is most dangerous and why.*

- *Define the toxicity measurements terms PEL, TLV, and STEL.*

- *List three diseases caused by blood-borne pathogens.*

INTRODUCTION

In the previous chapter, we studied health and physical hazards. In this chapter, we will study a new hazard called *toxic hazards. Toxicology* is the science that studies the harmful effects of chemicals on living tissue. A toxic (poisonous) substance is one that has a negative effect on the health of a human or animal. The harm or injury can be reversible or irreversible and may affect a single cell, an organ system or the entire organism. In other words, a toxic substance is going to hurt you, if not right away, then at some time in the future. Fear of toxic chemicals and their releases have increased because of:

- Industrial accidents such as at Bhopal, India
- Cargo vehicle ruptures (rail cars and tank trucks)
- Increased awareness of injuries to industrial plant workers
- Growing public awareness of toxic chemicals in the community

About every 20 minutes a new and potentially hazardous chemical is introduced into American industry. Highly reactive chemicals are used more extensively in the process industry and agriculture, and this has aroused concern and apprehension about their effects on the health of the surrounding communities. A major problem with toxic chemicals is identifying the toxic effects when their toxicity is not always obvious. It can take as long as 30 years before the adverse effects of exposure to a particular chemical becomes recognized. Smoking does not noticeably harm you right away, but 30 years of smoking will definitely show symptoms. This delay is known as *latency* and is a serious problem in identifying carcinogenic (cancer causing) compounds. Many years after being in common usage in a country some chemicals have been found to have insidious health effects and be long lasting in tissues and the environment. The chemicals dioxin and polychlorinated biphenyls are a prime example of environmental contaminants and carcinogens identified too late after being released into society and the environment. The best safeguard for the general public and industrial workers against toxic materials is knowledge and training.

All employees working on processing units that utilize toxic chemicals are at potential risks from toxic hazards. Many people think it is just the process operators who are at risk to hazardous chemicals. They fail to consider that instrumentation and analyzer technicians, and maintenance personnel are all at risk to the toxic chemicals on the process unit to which they are assigned and where they are working among the piping and vessels, changing out instruments, or replacing tubing that contains toxic chemicals.

This chapter also covers blood-borne pathogens, a source of real diseases and concern for everyone who becomes involved in emergency response and emergency medical care.

TOXIC SUBSTANCES

There are numerous government regulations addressing chemical hazards. Some of them, to name just a few, are:

- OSHA 29 CFR 1910.1200 Hazard Communication
- OSHA 29 CFR 1910.119 Process Safety Management
- OSHA 29 CFR 1910.120 Hazardous Waste Operations and Emergency Response
- EPA Toxic Substance Control Act
- EPA Resource Conservation and Recovery Act
- DOT hazardous Materials Handling: Loading and Unloading

A substance can be considered toxic (poisonous) when a small quantity will cause injurious effects in the body of the average normal adult human. Almost all materials are injurious to living organisms to some extent. Overdoses of table salt fed accidentally to infants have caused their deaths. Water, which we drink, and nitrogen and oxygen, which we breathe, can cause injury or death in overdoses. There is a difference between being exposed to a toxic substance and contaminated by a toxic substance. *Exposure* is when a chemical, infectious material or other agent enters or is in direct contact with the body. *Contamination* occurs when the hazardous material remains on the clothing, hair, skin, or other part of a person. Exposure can occur without contamination, but contamination usually results from exposure. Certain persons have unusual susceptibilities to substances that produce violent or fatal reactions when absorbed in small amounts harmless to most other persons. Such substances are called *allergens*, and a person is said to have an allergy or to be allergic to the substance. Allergies have been reported to almost every known common material.

Attempts have been made to determine the dosages of toxicants that cause injury to the normal (nonallergic) adult, but this has met with limited success. Generally, the greater the toxicity of the substance, the faster the rate of absorption, and the warmer the temperature, the more rapid the occurrence of the injury. In addition to individual susceptibility, other conditions affect the severity of the injury caused by the toxin. These other conditions are the:

- Size and duration of the dose
- Route taken into the body
- Degree of toxicity
- Rate of absorption
- Environmental temperature
- Physical condition of the affected person

MECHANISMS OF TOXIC AGENTS

Toxic agents cause injuries in different ways and have been categorized into seven types (see Figure 5-1): asphyxiates, irritants, systemic poisons, anesthetics, neurotics, corrosives, and carcinogens.

Asphyxiates

Although the term *asphyxia* is commonly thought to mean suffocation, it actually means *hypoxia* (lack of oxygen) and the presence of high carbon dioxide levels in the blood. Asphyxiates can be either "simple" or "chemical." Simple asphyxiates are generally considered to be those gases which dilute breathable air to such an extent that the blood receives an inadequate supply of oxygen. In other words, you are in an oxygen deficient atmosphere.

Some common asphyxiates are carbon dioxide and nitrogen. They are especially insidious because they are odorless and colorless. Hydrogen and helium are rarely asphyxiates because both are lighter than air and diffuse rapidly in the open. They tend to leak away rather than collect and displace air in an area. Nitrogen at room temperatures is slightly lighter than air, but at low temperatures, such as during leakage from tanks containing cryogenic nitrogen or where rapid expansion cools the gas, nitrogen becomes heavier than

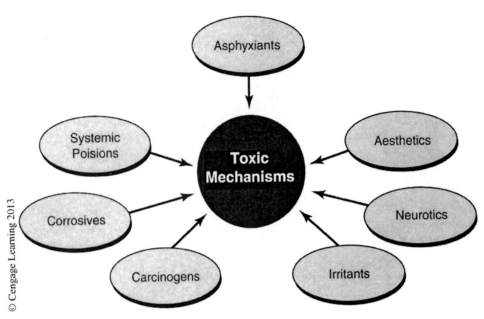

© Cengage Learning 2013

Figure 5-1 Mechanisms of Toxic Agents

the surrounding air. Cold nitrogen and carbon dioxide are all heavier than air and will settle in low spots such as sumps, the bottoms of tanks, or other low spots where unwary persons could be asphyxiated.

Chemical asphyxiates are toxic agents which enter into reactions to cause *histotoxic hypoxia*. These chemicals prevent the red blood cells from carrying oxygen. Some more familiar chemical asphyxiates are carbon monoxide, nitrites, hydrogen sulfide, and aniline.

Irritants

An irritant can be a gas, liquid, or fine particulate matter. Ammonia is a familiar and common irritant. Irritants injure the body by inflaming the tissues at the point of contact. Heat, redness, swelling, and pain are signs of inflammation. Mild irritants cause the capillaries to dilate and fill with blood, causing the redness and increased heat. The permeability of the capillary walls change, and fluid passes from the blood into the spaces between the tissues and causes swelling and pain. Strong irritants can produce blisters.

Particulate matter may be inert and nontoxic if ingested but its presence in the respiratory passages and lungs can be mechanically irritating. For example, asbestos is a nonreactive fibrous substance that severely irritates the respiratory system. It is a prime example of the toxic effects of particulate matter. *Time* magazine in January of 1974, observed that out of 869 persons employed for 17 years in an asbestos plant in Texas (which was closed in 1971), 300 workers "will die of asbestosis, a permanent and often progressive scarring of lung tissue from inhaled fibers, lung cancer or cancer of the colon, rectum or stomach."

The nasal passages contain the first defenses against harmful particulate matter entering the lungs. The hairs at the entrance act as a filter against larger particles, then the ciliated cells of the mucous membranes create a fluid flow that traps and washes out smaller particles. The body's defense mechanisms can be overwhelmed by massive and continued invasion of the respiratory system by fine solids. Except for those amounts removed by coughing, the particulate matter remains in the lungs. Long-time exposure to fine particulate matter has resulted in:

- *Black lung* disease from coal dust
- *Silicosis* from fine rock dust
- *Asbestosis* and cancer from asbestos fibers
- *Emphysema* from cigarette smoke

Dermatitis is an inflammation caused by defatting of the skin or by contact with an irritating or sensitizing substance. Exposure to solvents often removes oils that keep the skin soft and pliable. Without the oil, the skin is dry, scaly, and tends to crack easily. Such skin has poor resistance to bacterial infections and heals slowly when injured. A primary skin irritant causes dermatitis almost immediately by direct action on the skin. After a while, a person may become sensitized to a chemical. A sensitizer may not cause injury immediately, but will produce susceptibility to a second attack or to other substances. Where sensitivity or allergy dermatitis exists and the skin is affected even by small amounts of the chemical, the affected worker should not be allowed in areas where the chemical is used.

Systemic Poisons

Systemic poisons cause injury after they have been carried to the tissues of the body, especially to specific organs (target organs). Some systemics cause histotoxic hypoxia by interfering with the use of oxygen and others interfere with reactions necessary for the organs to continue their normal functions. For example, some toxic agents shut down the kidneys. The most damaging effects of systemic poisons occur at the kidneys, lungs, liver, and gastrointestinal tract. Systemic poisons can be divided into the following four categories:

1. Chemicals that cause injury to one or more of the visceral organs such as the kidneys or liver. The majority of the halogenated hydrocarbons belong in this group.
2. Chemicals that injure the bone marrow, spleen, and the blood-forming system and cause anemia and reduced count of white blood corpuscles. These toxicants include naphthalene, benzene, phenol, and toluene.
3. Chemicals that affect the nervous system and cause inflammation of the nerves. The result is tenderness of the nerves, pain, interference with transmission of nerve impulses, and even paralysis.
4. Toxic metals and nonmetals that not only cause respiratory system damage as irritants, but also can injure the body by being swallowed or entering the bloodstream through skin lacerations. These metals and nonmetals can be deposited in and interfere with functions of the body organs, bones, and blood. Generally, the effects of these substances are chronic in nature and take place only after continued and massive exposure.

Sometimes, metabolic changes to the chemical in the body create toxic products more injurious than the toxin originally absorbed. Because time is required for some substances to change within the body to other compounds, serious symptoms may not occur for several days after initial ingestion.

Some toxins may affect only specific organs called *target organs*. For instance, the toxins that affect the liver are called *hepatotoxins*, those affecting the kidneys are called *nephrotoxins*, those affecting the nervous system are called *neurotoxins*, and those affecting bone marrow and the blood system are called *hemotoxins*.

Carbon tetrachloride, trichloroethylene, and polyvinyl chloride (PVC) are well known hepatotoxins. Nephrotoxins are the chemicals that affect the kidneys. Since a large amount of blood circulates through the kidneys, they are susceptible to chemical attack if the blood contains a toxin. Heavy metals, chlorinated and fluorinated hydrocarbons, carbon disulfide, and ethylene glycol are some examples of kidney poisons. Many of these toxins can enter the body through inhalation or absorption. Toxins can be detected through medical surveillance, which monitors metabolites in the urine. Metabolites are the products formed by the toxin after it undergoes changes due to metabolism.

Reproductive toxins pose a danger to both male and female workers. They can cause gene mutations and affect the reproductive system by affecting sperm count in males or fertility in females. *Teratogens* are chemicals that affect the offspring. Polychlorinated biphenyls (PCBs), cadmium, and methyl mercury are known reproductive teratotoxins.

Anesthetics

Anesthetics cause the loss of sensation in the body. The loss of sensation can be general or local. Anesthetics cause respiratory failure by depressing the nervous system and interfering with involuntary muscular action. Some halogenated hydrocarbons generally used as cleaning or degreasing agents can produce this effect. Familiar compounds that have an anesthetic effect are ethyl ether, chloroform, and nitrous oxide.

Neurotics

Neurotics affect the nervous system, brain, or spinal cord and may be either depressants or stimulants. Ethyl alcohol and drugs are examples of neurotic agents that cause exhilaration and then deep depression.

A **depressant** is an agent that reduces functional activity and vital energies. Depressants may exhilarate for a short period of time and then cause the person to become drowsy and lethargic. Breathing may be labored and there may be a loss of consciousness. It can be selective and act on specific organs, such as the brain, while other organs are little affected. Drugs such as marijuana, cocaine, hashish, LSD, and others have adverse effects on the body (and safety). The effect of each depends on the type of drug, dosage, time since taken, susceptibility, and other factors.

Stimulants accelerate the function of an organ or system by exciting the affected person either slightly or severely. In extreme cases, the affected person may have a rapid pulse, jerking of the muscles, disorganized vision, and sometimes delirium. Caffeine, found in small amounts in coffee, stimulates the brain and nervous system. Stimulants taken to overcome fatigue wear off and the body succumbs to fatigue, perhaps at the wrong time, which could result in an accident. Because of scenarios like this, drugs are unacceptable in the workplace. They are responsible for erratic employee behavior.

Carcinogens

Certain chemicals cause cancer of the internal organs and systems of the body. Tar, bitumen, anthracene, and their compounds, products, and residues can cause cancers of the skin. Widely used chemicals that are carcinogens are vinyl chloride (liver cancer), benzene (leukemia), formaldehyde (nasal cancer), butadiene, ethyleneamine, trichloroethylene, and a long list of others. Many of these were found only after extensive worker complaints and analytical tests.

Corrosives

Corrosives damage by chemical destruction of the tissue they contact. Strong acids (hydrochloric, nitric and sulfuric) or alkalis (sodium hydroxide and potassium hydroxide) can cause corrosive burns. Alkalis can cause progressive burns, meaning the injury increases as the alkali moves through the damaged tissue. This is especially critical in injuries to the eye, where delicate tissues can be damaged little by little until vision is destroyed. The severity of a corrosive burn depends on:

- The concentration and type of corrosive chemical
- Whether the contact area was covered or uncovered by clothing
- The length of time of contact

Because of the bulleted reasons just mentioned, if contact is made with a corrosive, the corrosive should be washed away as soon as possible and neutralized with a mild antidote if one is available.

ROUTES OF EXPOSURE

A toxin may enter the body in many different ways but the four major routes are:

- Inhalation
- Ingestion
- Absorption
- Injection

Inhalation

Of these four routes, inhalation is the most common and most dangerous, since the toxin is brought inside the body and absorbed into the bloodstream and distributed throughout the body rapidly. Inhalation is the most prevalent source of exposure in the refining and petrochemical industry. Occupational Safety and Health Administration (OSHA) places high importance on this route because inhaled toxins reach body organs very fast and rapidly cause adverse effects. This route of exposure gives little time to administer corrective action. Inhalation sends the toxin to the blood stream and since the blood stream serves all the cells in the body, the toxin reaches all the cells. In many cases, the inhalation effects can be wide spread, affecting lungs, kidneys, bones, bone marrow, spleen, etc. Some examples of inhalation hazards are:

- Carbon dioxide, a simple asphyxiant, which dilutes oxygen and produces an hypoxic condition in the blood that starves the body cells of oxygen.
- Absorbed hexane vapors attack the peripheral nervous system and can cause peripheral neuropathy. The cumulative effect of hexane is to remove the protective sheath around the neurons. As the protective layer is eaten up, the nerves become damaged or destroyed and create a condition similar to paralysis.
- Many heavy metal ions, such as chromium or nickel, may damage the kidneys and are called *nephrotoxins*. They can be inhaled as a fine dust or mist. Though they damage the kidneys, their effect on other body organs may be minimal.

Figure 5-2 shows some major body systems susceptible to different toxins. Remember that different toxins affect different systems (target organs).

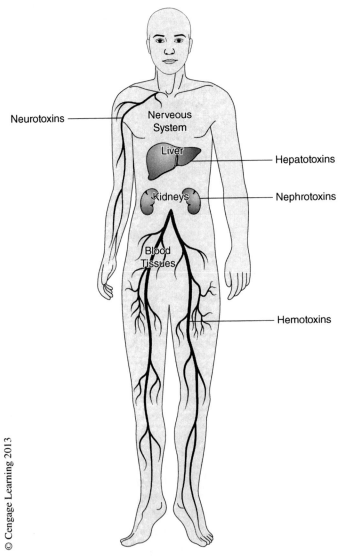

Figure 5-2   Body Target Organs of Different Toxins

Absorption

Next to inhalation, absorption is the second major route of entry of toxins into the body. Many liquids can be absorbed through the skin easily. Generally, lighter liquids are more likely to be absorption hazards than heavier or viscous (sticky) liquids. Phenol, a derivative of benzene, is an example of a chemical with a severe absorption hazard. Workers have died from contact with small amounts of liquid phenol, which is a neurotoxin.

Acids, alkalis, and phenols are acute toxins (the effect appears rapidly). Acids and alkalis destroy the skin tissue while phenol rapidly attacks internal body organs. Other absorbed materials can affect the body in a chronic way (the effect becomes visible only after repeated exposures). Personal protective equipment (PPE) is the most widely used means of protection against absorption. Gloves, coveralls, and face shields protect the worker from splash contact with toxic chemicals. However, PPE, if it is not donned properly, is useless for protection.

If a person becomes splashed with a large amount of a toxin, keep in mind the clothing of the exposed person can cause harm to emergency response personnel. The toxin can be absorbed from the contaminated clothes to the clothing or skin of those rendering aid. The best practice is to undress the contaminated person, put them under a shower, isolate their contaminated clothing, and then transport the person to other locations or a hospital. Some companies have a standard policy that any person exposed to skin absorptive toxins will undergo a medical checkup immediately. Contaminated clothing should never be taken home for cleaning because this risks spreading the toxin throughout a household. Leather absorbs and holds chemicals well, so leather clothing should be discarded.

Another important route by which toxic agents can enter the bloodstream is through the gastrointestinal system. Entry of the poison into the digestive system might occur when a toxic substance is ingested when contaminated fingers, food, or other objects are placed in the mouth. Washing the hands prior to eating is an important means of preventing accidental poisoning. Most sites have strict lunchroom rules about leaving PPE and contaminated clothing outside the lunchroom. Another route into the gastrointestinal tract is the respiratory system. After exposure to a toxin, the respiratory system may reject toxic particulate matter by clearing the throat or coughing. If this toxic matter is swallowed, it enters the gastrointestinal tract. Solid particles filtered out by the cilia are moved up slowly to the mouth. Swallowing these accumulations permits absorption of the toxicant through the digestive system.

Injection and Ingestion

Generally, injection and ingestion are relatively uncommon in industry. An injection is a cut, puncture, or laceration that allows a toxin direct entry into the blood stream. Naturally, the effect of the toxin under these circumstances could be fast and intense. Ingestion, the swallowing of a toxin, is very rare but can happen. Most commonly it occurs if a person falls into a vessel full of a liquid or powder, or a line under pressure ruptures and the liquid strikes the worker in the face while they are talking (the mouth is open).

MEASUREMENT OF TOXICITY

Often an association can be made between a chemical hazard and an occupational illness or injury. This requires an understanding of the *dose-response relationship*. A dose-response relationship means that an increasing effect can be correlated with an increasing dose of toxin. Dose-response relationships hold for most body systems except the immune system.

Dose-response is statistical in nature. A particular person's response may differ significantly from that predicted by dose-response. Dose-response is a guideline and not a direct enforcement tool. Dose-response attempts to capture the toxicity in a simple manner, but recognizes that toxicity is not a straightforward indicator. Experts recognize this and realize toxicity can be influenced by a variety of factors including age, gender, length of exposure, ambient conditions (humidity, temperature, etc.), age, and many others. There is no one system to measure toxicity. There are several and we will discuss some of the more important ones.

PEL, TLV, and STEL

The table of *threshold limit values (TLVs)* has been published annually by the Threshold Limit Values committee of the American Conference of Industrial Hygienists (ACGIH). The TLV committee reviews and updates the specific recommended exposure limits for each substance as new information is obtained. The TLV list is not only used in the United States, but is often adopted verbatim in other countries. In the OSHA Act of 1970, the TLVs of 1968 were adopted and given the status of law. OSHA named them the *permissible exposure limits (PELs)*.

Both PEL and TLV refer to toxic chemical exposure levels. Just remember, the PEL values of toxic chemicals are published by OSHA, while the TLV values are published by the ACGIH. You can find PEL values on the OSHA Web site (www.osha.gov). Look under *regulations*. PEL values refer to inhalation exposure. TLV values are published by ACGIH annually. Most chemicals have identical PEL and TLV values.

Some people consider the TLVs and PELs as being levels at which exposed personnel are not at risk. This is not true. TLVs and PELs have been set as guidelines for control of the workplace atmosphere. They are primarily for toxic agents that enter the body through the respiratory system. It is always good practice to hold any exposure to the minimum practically possible. For instance, hexane has a PEL of 50 parts per million (PPM). What does 50 PPM mean? Simply put, no exposure of 50 PPM (average for an eight-hour period) is permissible. Continuous and chronic exposure above 50 PPM may have obvious adverse health effects. How do we use this information in practical situations? At a very basic level, workers are not to be exposed to an average of 50 PPM of hexane vapor. The key word is *average*. For example, in an eight-hour day, an employee may be exposed to more than the PEL value, but the average for eight hours should be at or below 50 PPM. Since process operators work 12 hour shifts the PEL values are adjusted to approximately (2/3 or 67 percent) of their published values. For hexane, the published PEL is 50 PPM, but the effective value for a 12-hour exposure is approximately 33.5 PPM. This adjustment is known as the Box-Scala adjustment. In some cases, an extremely high instantaneous exposure level which will average at or below the PEL value may be harmful.

PEL and TLV are good exposure guidelines. The distinction between PEL and TLV is in terms of legal impact. PEL values are enforceable by OSHA, while the TLV values are guides and have no enforcement capability associated with them. PEL and TLV are average exposures. For a given instant, you can have exposure higher than 50 PPM as long as the average is at or below the PEL.

For some chemicals, the material safety data sheet Material Safety Data Sheet (MSDS) show a short-term exposure limit (STEL) value. STEL is defined as the 15-minute average that should not be exceeded even if the average value for an eight-hour day is below the PEL. STEL values are related to short term effects while PEL values apply to chronic exposures. The following are some definitions used by OSHA:

Time Weighted Average (TWA) is the employee's average airborne exposure in any eight-hour work shift of a 40-hour workweek which shall not be exceeded.

Permissible Exposure Limits (PEL) are limits to indicate the maximum airborne concentration of a contaminant to which an employee may be exposed over the duration specified by the type of PEL assigned to that contaminant.

Short Term Exposure Limit (STEL) is the maximum airborne concentration which workers can be exposed to for periods up to 15 minutes continuously without suffering from irritation, or chronic tissue changes, or narcosis, provided there are no more than four excursions per day with at least 60 minutes between exposure periods.

Ceiling (C) is a concentration that the employee's exposure shall not be exceed during any part of the work day. If instantaneous monitoring is not feasible, then the ceiling shall be assessed as a five-minute time weighted average exposure which shall not be exceeded at any time over a working day. Some chemicals have ceiling concentrations values that should never be exceeded.

Lethal Dose 50 (LD_{50}) is the dose that causes death for 50 percent of the test subjects. Since dead test subjects are obvious, the LD_{50} provides good information in a given species to the toxicity of a given toxin. Lethal doses do not apply to inhaled substances. This term is used for solids or liquids that are absorbed through the skin. Dosage is expressed in milligrams per toxicant per kilogram of body weight of the test animal used. Extrapolating from that, a rough approximation of the dosage that would affect humans can be calculated.

Lethal Concentration 50 (LC_{50}) is the airborne concentration of a substance which is lethal to 50 percent of the test animals. The term is used for inhalation hazards.

 BLOOD-BORNE PATHOGENS

The health hazards of blood-borne pathogens is an important concern of medical and emergency response personnel, and not just those who work for municipalities, state, or local agencies. Most processing industries have their own medical and emergency response personnel. When workers receive a serious trauma that results in a bleeding wound of some type, responders are at risk to blood-borne pathogens. Acquired Immune Deficiency Syndrome (AIDS) is only one of about a dozen diseases caused by blood-borne pathogens. Other common ones are hepatitis B, syphilis, brucellosis, and malaria. Because of the risk associated with occupational exposure to blood-borne diseases, OSHA passed the Blood-borne Pathogen Standard 29 CFR 1901.1030 in 1992.

Many emergency response personnel and workers applying first aid may be exposed to blood-borne pathogens through eye, mouth, skin, or mucous membrane contact with infected body fluids of an injured person. Contamination occurs mostly from exposure to small sprays, splashes, or mists of body fluids. Because most of these contaminations do not cause an immediate adverse health effect, many workers do not truly understand the hazards they are exposed to when responding to an injured employee. Exposure can occur through broken skin allowing infectious entry while rendering assistance or during cleanup of contaminated material.

Emergency service personnel can also be exposed to blood-borne pathogens in a similar manner when treating injuries, performing cardiopulmonary resuscitation, and carrying victims. It is not possible to tell simply by looking at a person whether the person is infected with HIV, hepatitis, or other blood-borne diseases. This and other observations led to recommendations for universal precautions at all times. The concept behind universal precautions is that all human blood and body fluids are treated as if they are known to contain blood-borne pathogens. All employees should take this approach whenever they respond to a bleeding injured coworker. They should avoid skin contact with blood, body fluids, or other potentially infectious materials (such as bloody clothing).

The OSHA Blood-Borne Pathogen Standard mandates that each facility must establish its own blood-borne exposure control plan to avoid placing employees in contact with blood, body fluids, or other potentially infectious materials. Each facility develops its own program and trains employees to the requirements of that program.

SUMMARY

A typical petrochemical plant or a refinery has hundreds of chemical hazards. Lack of care in handling or storage of the chemicals can cause bodily harm that can vary in intensity from minor to severe or even death. However, due to effective standard operating procedures, engineering controls, and personal protective equipment these risks have been controlled relatively effectively. Training, procedures, engineering controls, and administrative procedures are collectively responsible for this success.

Toxicology is the science that studies the harmful effects of chemicals on living tissue. A toxic (poisonous) substance is one that has a negative effect on the health of a human or animal. The harm or injury can be reversible or irreversible and may affect a single cell, an organ system or the entire organism. Generally, the greater the toxicity of the substance, the faster the rate of absorption, and the warmer the temperature the more rapid the occurrence of the injury.

Toxic mechanisms allow toxic agents to cause injuries in different ways. Toxic mechanisms have been categorized into seven types: asphyxiantes, irritants, systemic poisons, anesthetics, neurotics, corrosives, and carcinogens. A toxin may enter the body in many different ways but the four major routes are inhalation, ingestion, absorption, and injection. Of these, inhalation is the most common and most dangerous, since the toxin is brought inside the body and absorbed into the bloodstream and distributed throughout the body rapidly.

The health hazards of blood-borne pathogens—AIDS, hepatitis B, syphilis, etc.—are an important concern of medical and emergency response personnel. When workers receive a serious trauma that results in a bleeding wound of some type, responders are at risk to blood-borne pathogens. Many emergency response personnel and workers applying first aid may be exposed to blood-borne pathogens through eye, mouth, skin, or mucous membrane contact with the infected body fluids of an injured person.

REVIEW QUESTIONS

1. _____ is the science that studies the harmful effects of chemicals on living tissue.

2. It can take as long as 30 years before the adverse effects of exposure to a particular chemical becomes recognized. This delay is known as _____.

3. Carcinogenic compounds cause:
   a. Heart problems
   b. Liver problems
   c. aneurysms
   d. cancer

4. (T/F) Like process operators, analyzer technicians, and instrumentation technicians are all at risk to toxic chemicals.

5. A substance can be considered _____ when a small quantity will cause injurious effects in the body of an average adult human.
   a. hazardous
   b. corrosive
   c. toxic
   d. carcinogenic

6. _____ is when a chemical, infectious material, or other agent enters or is in direct contact with the body.

7. _____ occurs when the hazardous material remains on the clothing, hair, or skin.

8. List four mechanisms by which toxic substances cause bodily injury.

9. _____ affect the body by inflaming the tissues at the point of contact.

10. Explain how corrosives affect the body.

11. _____ injure specific parts of the body called target organs.

12. List the four routes of entry of toxic substances into the body.

13. The most dangerous route into the body is by _____.

14. _____ toxins attack the bone marrow and blood system.

15. _____ toxins attack the nervous system.

16. (T/F) Inhalation is so dangerous an entry because it happens so fast.

17. Explain the term dose-response relationship.

18. (T/F) PEL values refer to inhalation exposure.

19. Which of the two, PEL or TLV, is enforced by a standard?

20. _____ is the airborne concentration of a substance that is lethal to 50 percent of the test animals.

21. List three diseases caused by blood-borne pathogens.

22. (T/F) Contamination from blood-borne pathogens occurs from exposure to small sprays, splashes, or mists of body fluids from infected people.

EXERCISES

1. Research on the Internet the combination of the words (latency) + (toxins) and write a two page report of your findings of one or two chemicals.

2. Research on the Internet hazardous nature of the strong corrosive sodium hydroxide, also called *caustic*. This is a very common chemical used in the processing industry. Write a one page report.

RESOURCES

www.wikipedia.org

www.osha.gov

www.YouTube.com

www.cdc.gov

www.cdc.gov/niosh/

www.toolboxtopics.com

CHAPTER 6

Fire and Fire Hazards

Learning Objectives

Upon completion of this chapter the student should be able to:

- *Explain why fire prevention is important to the processing industry.*

- *Describe the four factors required for a fire to occur and keep burning.*

- *Describe the four classes of fires.*

- *List four products of combustion.*

- *Describe three systems used to detect fire hazards.*

- *Explain how isolation can be used to suppress fires.*

- *Describe five types of firefighting equipment.*

INTRODUCTION

Opportunities for fires in petrochemicals, refining, manufacturing, and oil and gas production abound. Fires are just another hazard employees must guard against. There are tank farms of numerous large and small storage tanks, spherical liquefied natural gas storage tanks, pipelines full of flammables, not to mention reactors, distillation towers, drums, etc. Fuel is everywhere in enormous quantities around the working employees. Air is everywhere around vessels containing fuels and flammables. All that is needed is a loss of containment and the right air-fuel mixture combined with a source of ignition. The potential in industry for loss of life or injury is serious enough that the Occupational Safety and Health Administration (OSHA) has mandated certain standards for fire protection in 29 CFR 1910.155-165.

Fire and explosion accidents are of major concern to the owners and operators of refineries and petrochemical, gas processing, terminal, and offshore facilities. Statistics have shown that the majority of monetary loss in these types of complexes is due to fire and explosion. According to statistics (www.ohsonline.com, December 2010), 77 percent of the monetary loss in refinery and petrochemical complexes is due to fire and explosion. The breakout of accidents due to fire and explosion is 65 percent vessel (container) and vapor cloud explosion and 35 percent fire. The causes of these accidents are mostly attributed to mechanical issues, process upset, and operator error.

Corporations have an enormous investment in their processing plants. Even fires that did not obviously appear to have damaged vessels, pipes, or pumps may have altered the integrity (metallurgy) of the equipment so that they must be replaced at great expense. Industry spends large sums of money on fire detection, firefighting equipment, and fire fighter training (see Figure 6-1). Training and equipment include fire fighting drills, fire trucks, fire monitors and turrets, fire brigades, bunker gear, and large numbers of locally mounted fire extinguishers. Fire prevention and fire fighting are critical issues with processing industries.

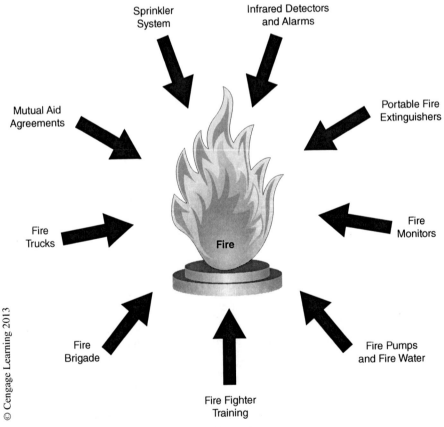

Figure 6-1 Industry Investment in Fire Prevention

© Cengage Learning 2013

FIRE HAZARDS

Once fuel and an oxidizer (air) combine in the right proportions (not too rich, not too lean) and a source of ignition is found, fire occurs. Fire is a rapid chemical reaction that gives off energy (light and heat) and products of combustion that are very different from the fuel and oxygen that combined to produce the products. Many of the gas products are toxic, some highly toxic. The majority of people that die in fires rarely burn to death. They are asphyxiated from breathing smoke or poisoned by toxic fumes from burning plastics, carpets, paints, etc. Carbon monoxide is the number one killer in fires and is produced in almost all burning organic compounds. During fires carbon monoxide is produced in large quantities and can quickly reach lethal concentrations. Fires burning with a lack of oxygen instead of plentiful oxygen produce a significantly wider amount of compounds, many of them very toxic. Partial oxidation of carbon produces carbon monoxide, nitrogen-containing materials can yield hydrogen cyanide, ammonia, and nitrogen oxides. Halogen containing compounds may combust into hydrogen chloride, phosgene, and dioxins, all extremely deadly.

Fire hazards are conditions that favor fire development or growth. Three substances are required to start a fire and one to sustain fire: (1) oxygen, (2) fuel, and (3) and a source of ignition to start the fire, and heat to sustain the fire. Since oxygen is normally present almost everywhere on earth, the creation of a fire hazard usually involves the mishandling of fuel or heat. Fire (combustion) is a chemical reaction between oxygen and a combustible fuel. **Combustion** is the process by which fire converts fuel and oxygen into energy, usually in the form of heat. By-products of combustion include light, heat and smoke (gases and fumes). For a fire to start, a source of ignition, such as a spark, open flame or a sufficiently high temperature is needed. Given a high enough temperature, almost every substance will burn. The **combustion point** is the temperature at which a given fuel can burst into flame.

Fire is a chain reaction that requires a constant source of fuel, oxygen, and heat. Exothermic chemical reactions generate heat. Combustion and fire are exothermic reactions and can generate large quantities of heat. An ongoing fire usually provides its own sources of heat. It is important to remember cooling is one of the principal ways to control a fire or put it out because for it to continue burning it must continue to have heat.

Liquids and solids, such as oil and wood, do not burn directly but are first converted into a flammable vapor by heat. Hold a match to a sheet of paper and the paper will burst into flames. Look closely at the paper and you will see that the paper is not burning. The flames reside in a vapor area just above the surface of the sheet. Vapors will only burn at a specific range of mixtures of oxygen and fuel, determined by the composition of the fuel. Remove a fire's access to fuel or remove the oxygen, and the fire dies. Although a spark, flame, or heat may start a fire, the heat that a fire produces is necessary to keep it going. Therefore, a fire may be put out by removing the fuel source, starving it of oxygen, or cooling it below its combustion point.

Sources of Fire Hazards

Almost everything in an industrial environment will burn. Metal furniture, machines, plaster, and concrete block walls are usually painted. Most paints and lacquers will easily burn. Therefore, the principal method of fire suppression is passive—the prevention (absence)

Table 6-1 Classes of Fire

Classes of Fires	
Class A	Solid materials such as wood, paper, plastic, housing, etc.
Class B	Flammable liquids and gases
Class C	Electrical (within breaker boxes, motor control centers, etc.)
Class D	Combustible metals, such as aluminum, magnesium, and titanium

© Cengage Learning 2013

of a source of ignition. Within our daily home and work environment various conditions elevate the risk of fire and so are termed *fire hazards*. Examples are a garage containing a five-gallon can of gasoline and several cylinders of propane fuel for lanterns, or a storage room at work full of boxes of paper or other combustibles. For identification purposes fires are classified according to their properties, which relate to the nature of the fuel. The properties of the fuel directly correspond to the best means of combating a fire. This is revealed in Table 6-1.

Without a source of fuel, there is no fire hazard. Fuels occur as solids, liquids, vapors, and gases. Solid fuels include wood, building decorations, and furnishings such as fabric curtains and wall coverings, and synthetics used in furniture. What would an office be without paper? What would most factories be without cardboard and packing materials like Styrofoam™ molds, shredded papers, and solvents? All of these materials easily burn. We can make many things fire retardant but few solid fuels can be made fireproof. Wood and textile can be treated with fire or flame-retardant chemicals to reduce their flammability but under the right conditions they will still burn. Solid fuels are involved in most industrial fires, but the presence of flammable liquids and gases is a major cause of industrial fires.

Two terms often confused and applied to flammable liquids are **flash point** and **fire point**. The flash point is the temperature for a given fuel at which vapors are produced in sufficient concentration to flash in the presence of a source of ignition. The fire point is the minimum temperature at which the vapors will continue to burn given a source of ignition. Flammable liquids have been defined as having a flash point below 100°F. Combustible liquids have been defined as having a flash point at or higher than 100°F. These are further divided into the three classifications based on flash points. As the temperature of any flammable liquid increases, the amount of vapor generated on its surface also increases. Safe handling, therefore, requires both knowledge of the properties of the liquid and an awareness of ambient temperatures in the work or storage place.

Here are a few more interesting facts about fuels. Most flammable liquids are lighter than water. If they catch fire, water cannot be used to put such a fire out. Water floats the burning fuel and spreads it wherever the water flows. We have all seen this in naval warfare movies of sinking ships and bunker fuel oil burning on top of water. Many gases are lighter than air. Released into air, gas concentrations are difficult to monitor due to the changing factors of air, current direction, and temperature. Gases may also stratify in layers of differing concentrations but lighter ones will often collect near the top of whatever container in which they are enclosed. Concentrations may have been sampled as being safe at chest level but a few feet above the head may exceed flammability limits.

Reducing Fire Hazards

In petrochemical plants and refineries where fuel is everywhere and the plants are surrounded in an envelope of air, the best way to prevent fires is to eliminate the sources of ignition. Several ignition sources can be eliminated or isolated from fuels through administrative or engineering controls, such as:

- Smoking should be prohibited near any possible fuels.
- Electrical sparks from equipment, wiring, or lightning should not be close to fuels or areas containing fuels. Intrinsic safety features alleviate this problem.
- Open flames (welding torches, heating elements, etc.) should be kept separate from fuels and allowed only through a permit system.
- Tools or equipment that may produce mechanical or static sparks should be isolated from fuels.

FIREFIGHTERS

Because fire is a very serious hazard in the petrochemical and refining industries, a job called fire watch was created for special conditions, principally during unit turnarounds. During turnarounds a lot of hot work (welding, grinding, use of gasoline engines, etc.) occurs. The hot work provides a source of ignition for the flammables and combustibles in the plant. A technician assigned to fire watch, surveys work in a certain area, keeps the area clear of combustible material, maintains permit conditions, prevents actions that would result in the release of flammable substances, and sounds the alarm in the event of a fire.

Petrochemical and refining sites have to be prepared to fight a fire should one occur. They usually have their own firefighters, fire trucks, and firefighting gear. They will have a sufficient number of personnel trained as firefighters so that day or night, which ever crews are working, should the alarm sound, enough firefighters will be available. Depending on the size of the facility, two or more fire trucks may be available. The site will also have a mutual aid agreement with nearby plants to assist each other in case of fire. A neighboring plant will send its firefighters and fire trucks over to help its neighbor.

Firefighters have regularly scheduled drills where firefighters respond to a fictional fire on a specific unit. The fire trucks roll out, hoses are unloaded and rolled to the nearest fire hydrant, connected, and water sprayed on the fictional fire. Fire monitors are also operated. At most plants, firefighters are sent on a regular schedule, such as once a year, to a firefighting school (see Figure 6-2), where they get very intensive training and are brought up to date on advances in firefighting techniques.

FIRE DETECTION, SUPPRESSION, AND FIGHTING SYSTEMS

Many plants use a combination of systems to detect, suppress, and/or fight fires. Many of the systems, besides alarming locally at the site of the fire, also alarm to a remote location such as the site's emergency management office or guard building. Each system will be discussed.

© Cengage Learning 2013

Figure 6-2 Firefighter Training

Fire Detection Systems

Automatic fire detection systems are used in most industries today. They warn of the presence of smoke, radiation, elevated temperature, or increased light intensity. Some detector types are:

- Thermal expansion detectors that use a heat-sensitive metal link that melts at a predetermined temperature to make contact and sound an alarm.
- Photoelectric fire sensors that detect changes in infrared energy radiated by smoke or by the smoke particles obscuring the photoelectric beam. A relay closes to complete the alarm circuit when smoke interferes with the intensity of the photoelectric beam.
- Ionization or radiation sensors that use the tendency of a radioactive substance to ionize when exposed to smoke. The substance becomes electrically conductive with the smoke exposure and permits the alarm circuit to be completed.
- Ultraviolet or infrared detectors (see Figure 6-3) that sound an alarm when the radiation from flames is detected.

© Cengage Learning 2013

Figure 6-3 Infrared Fire Eye

Fire Suppression

When we suppress fires we inhibit their growth and spread. There are various means of fighting and suppressing fires. We will look at five methods of suppressing fires which are isolation, water, gas extinguishants, foams, and solid extinguishants.

Isolation of fuel from the oxidizer can be accomplished in several ways. If the fire is fed by fuel from a ruptured line, an upstream valve can be closed. This isolates the source of fuel. Blanketing with an inert gas or foam isolates the fuel from air.

Water is the most common fire extinguishant because it is generally available, low in cost, and effective. It is most commonly used on Class A combustibles. The principle effect of a stream of water is to cool the burning fuel below its ignition temperature. The water can be applied in a stream with the advantage of fire fighters being a good distance from the fire. Water applied as a spray or fog acts also as a coolant. A spray or fog cools by absorbing heat of vaporization and by breaking chain reactions. Water applied in this manner will absorb enough heat to vaporize and becomes steam, which as a gas above the fire, reduces the concentration of oxygen and thus further lowers the rate of combustion.

Gas extinguishants are not used as commonly as water but in some cases are more effective and the preferred method if the fire is in an enclosed space and the equipment in the room is very valuable (computers, motor control centers, etc,). There is no water damage or dry powder damage to the equipment or messy cleanup.

- Carbon dioxide (see Figure 6-4) is the most common gas extinguishant. It acts as a coolant, a blanketing agent, by reducing oxygen levels, and as a combustion inhibitor.

© Cengage Learning 2013

Figure 6-4 Carbon Dioxide Cylinders for Fire Suppression

- Nitrogen acts as a diluent and reduces the concentration of oxygen in air to less than that required for continued combustion.
- Halogenated hydrocarbons (CBr_2F_2, $CHBrCl$, CBF_3, etc.) act by inhibiting chain reactions. The disadvantage of the halogenated hydrocarbons is that even in unreacted states they are highly toxic and damaging to the ozone layer.

Foams suppress fires by cooling and blanketing and sealing off the burning fuel from the surrounding atmosphere. Foams gradually break down under the action of heat or water. Foams cannot be used with fires involving gaseous fuels, such as propane or materials that react violently with water.

Solid extinguishants are mainly used for combating Class B fires, such as oil or grease. The advantage of a solid extinguishant is its ability to remain on a burning surface without flowing off. Sodium and potassium bicarbonate are the principal solid extinguishants used on liquid fuels. Both solids decompose in water into carbon dioxide and salts. Also, both interfere with the chain reactions involving carbon that promotes combustion.

TYPES OF FIRE FIGHTING EQUIPMENT

Fire-fighting equipment can be portable, mobile, or fixed, and may be manual or automatic. Because of the prime importance of operational fire-fighting equipment, strict equipment inspection schedules are adhered to and recorded. Defective equipment is immediately replaced. Inspection of fire and safety equipment is normally carried out by process employees as part of their scheduled routines. Firefighting equipment that would be on their checklist include:

- Fire extinguishers (hand-held and mobile)
- Fire monitors and turret nozzles (grade and elevated)
- Fire hose stations
- Fire hose carts
- SCBA stations

Additional firefighting equipment in processing industries include:

- Sprinkler systems
- Halon systems
- Hand-held portable fire extinguishers
- Mobile fire extinguishers

Sprinkler and Halon systems are usually ceiling-mounted, fixed fire-fighting systems. One releases a deluge of water, the other a suppressive gas. Halon systems, which use a gas with the vendor name of Halon, are being phased out because Halon gases are damaging to the ozone layer.

Portable fire extinguishers may be filled with water, dry powder, foam, or carbon dioxide. They are usable from 8–120 seconds, depending on the size of the extinguisher and the chemical inside. The proper fire extinguisher should be placed in the proper place at a manufacturing site. If a fire breaks out, employees should not have to go a great distance to get an extinguisher nor have to determine if they have the correct type of extinguisher for the fire. The extinguisher should have marked on the side what type of fires they may be used on. Personnel should be trained annually (an OSHA requirement) on how to use the extinguisher and the proper way to fight the fire. OSHA has mandated the monthly and annual inspection and recording of the condition of fire extinguishers in industrial settings. A hydrostatic test to determine the integrity of the fire extinguisher metal shell is recommended according to the type of fire extinguisher. The hydrostatic test measures the capability of the shell to contain internal pressures and the pressure shifts expected to be encountered during a fire. Extinguishers should be located near exits and fire hazards. Process employees routinely inspect all fire extinguishers for operability.

Mobile fire extinguishers usually are mounted on a two-wheeled cart. They look like a hand-held fire extinguisher but are much larger, too large for a man to lift. One or more are strapped to a cart and can be rolled to the scene of a fire. They have a longer and larger fire hose than a hand-held extinguisher. They may be filled with a variety of fire extinguishants.

Fire monitors (see Figure 6-5) and turrets are special devices for throwing large streams of water for either long distances or high elevations. The monitor steadies the nozzle and holds the stream on target. After the stream has been set on target, the nozzle can be clamped in position and the monitor can be left unattended or an operator may continue to operate it. Because monitors can use large nozzle tips and operate at high pressures, they supply large volumes of cooling water. Monitors may be either fixed or portable. Monitors mounted on water mains around operating units are fixed. The operator can quickly turn on a monitor, aim and lock the stream on target, then leave it unattended while they perform other emergency duties. As a rule, the fixed monitor is used as an aid in controlling a fire. The portable monitor can be carried or moved from one location to another. It is mounted on a plate (or may have feet) and is connected to a fire hose. It may be manned or left unattended.

Figure 6-5 Fire Monitor

SUMMARY

Opportunities for fires in the processing industries abound. Fire is a chain reaction that requires a constant source of fuel, oxygen, and heat. The majority of people that die in fires are asphyxiated from breathing smoke or poisoned by toxic fumes. Carbon monoxide is the number one killer and is produced in almost all burning organic compounds. During fires, carbon monoxide is produced in large quantities and can quickly reach lethal concentrations.

Fire hazards are conditions that favor fire development or growth. Since oxygen is normally present almost everywhere on earth, the creation of a fire hazard usually involves the mishandling of fuel or heat. Automatic fire detection systems are used in most industries today. They can warn of the presence of smoke, radiation, elevated temperature, or increased light intensity.

When we suppress fires, we inhibit their growth and spread. There are various means of fighting and suppressing fires. Some of them are isolation, foam, water, and gas and solid extinguishants. Common firefighting equipment used in the processing industries are sprinkler systems, Halon systems, hand-held portable fire extinguishers, mobile fire extinguishers, and fire monitors and turrets.

REVIEW QUESTIONS

1. Explain why fire prevention is important to the processing industry.

2. List the three factors required for a fire to occur are:
   a. Fuel
   b. Combustibility
   c. Flame point
   d. Air
   e. Source of ignition

3. _____ kills the majority of people caught in a fire.

4. The byproducts of combustion are _____, _____, and _____.

Match the following three sentences from the alphabetical list below them.

5. Class B fire _____.

6. Class C fire _____.

7. Class A fire _____.
   a. Electrical
   b. Flammable liquids and gases
   c. Wood, paper, plastic, etc.

8. Explain why water cannot be used to fight a fire on a substance that is lighter than water.

9. Automatic fire detection systems warn of the presence of _____, _____, and _____.

10. (T/F) Hot work can provide a source of ignition for a flammable atmosphere.

11. (T/F) Site firefighters attend firefighting schools on a regular basis.

12. _____ separates a fuel from its oxidizer.

13. _____ is the most common fire extinguishant because it is generally available, low in cost, and effective.

14. _____ suppress fires by cooling, blanketing, and sealing off the burning fuel from the atmosphere.

15. The most common gas extinguishant is _____.

16. List five types of firefighting equipment.

17. (T/F) All portable fire extinguishers are filled with the same firefighting material.

18. Explain how fire monitors are used to fight fires.

EXERCISES

1. Go to www.YouTube.com and find a video about fire monitors, foam used in fire fighting, or the toxicity of smoke. Write a one page report about what you discovered. Cite your sources.

2. Research the Internet and find a PowerPoint presentation about fire detection systems or the toxicity of smoke. Download the presentation to a flash drive or CD and bring it to class to be presented by yourself or the instructor. Cite your sources.

RESOURCES

www.wikipedia.org

www.osha.gov

www.YouTube.com

www.toolboxtopics.com

CHAPTER 7

Hazards of Pressure, Steam, and Electricity

Learning Objectives

Upon completion of this chapter the student should be able to:

■ *List four types of hazards associated with high-pressure systems.*

■ *Describe two ways boilers are protected from over pressurization.*

■ *Describe four ways to detect gas leaks.*

■ *Describe four electrical hazards.*

■ *Explain the difference between* bonding *and* grounding.

■ *Explain how equipment is made explosion proof.*

INTRODUCTION

On April 27, 1865, the side-wheeler steamer *Sultana,* overloaded and carrying more passengers than it should, steamed up the Mississippi River. The *Sultana* carried more than 2,000 Union soldiers, many bound for home after being released from Confederate prison camps. Quick repairs had been made to the vessel's boilers at Memphis. A few miles north of Memphis, the boilers blew up and tore the *Sultana* apart, hurling men and parts of the vessels hundreds of feet. An estimated 1,700 soldiers died either from the explosion or from drowning. The pressures at which the *Sultana's* boiler normally operated, and even the pressure at which it ruptured so violently, would be considered low compared with boiler pressures commonly used today.

It is not necessary to have much pressure to create conditions where serious injuries and damage can occur. It is commonly and mistakenly believed that injury and damage will result only from high pressures, however, there is no agreement on the definition of the term *high pressure* beyond the fact that it is greater than normal atmospheric pressure.

- The American Gas Association states that a high-pressure gas distribution line is one which operates at a pressure of more than 2 pounds per square inch (psia).
- The American Society of Mechanical Engineers (ASME) rates only those boilers which operate at more than 15 psi as high-pressure boilers.
- OSHA standards state: "High-pressure cylinders mean those with a marked service pressure of 900 psi or greater."
- The military services and related industries have categorized low pressure to be from 14.7 to 500 psia; medium pressure from 500 to 3,000 psia; high pressure from 2,000 to 10,000 psia; ultrahigh pressure to be above 10,000 psia.

Thus, *high pressure* can be almost any level prescribed for the equipment or system in use. For accident prevention purposes, any pressure system must be regarded as hazardous. Hazards lie both in the pressure level and in the total energy involved. Any employee on a processing unit—operator, instrument and analyzer technicians, mechanical personnel, contractors—that has pressurized lines or vessels is subject to the hazards of pressure.

This chapter introduces workers in various industries to several more hazards they must guard against: high pressure, steam, and electricity. Remember we said in Chapter 1 that risk was everywhere? Maybe we should have added that multiple risks are everywhere. All work sites have multiple risks, from heights (stairs), electricity (office equipment), punctures (stapler, scissors, etc.), ergonomics, and many more.

HAZARDS OF PRESSURE

Pressure is defined as (1) the force exerted against an opposing fluid, or (2) force distributed over a surface. Pressure can be expressed in force or weight per unit of area, such as pounds per square inch (psi). Critical injury and damage can occur with relatively little pressure. Employees in processing industries are often surrounded with pipes and vessels under pressure. The hazards most commonly associated with high-pressure systems are leaks, pulsation, vibration, release of high-pressure gases, and whiplash from broken high-pressure tubing and hoses. Some strategies for reducing high-pressure hazards include:

- Strict adherence to design codes for vessels and piping systems
- Limiting vibration through the use of vibration dampening
- Overpressure relief systems
- Engineering controls

Rupture of Pressurized Vessels

When the pressure of a fluid inside a vessel exceeds the vessel's strength, it will fail by rupture. A slow rupture may occur by popping rivets or by opening a crack. If the rupture is rapid, the vessel can literally explode, generating metal fragments and a shock wave with blast effects as damaging as those of exploding bombs. Boiler explosions are often disastrous.

A boiler rupture occurs if steam flow output is prevented or restricted and the temperature and pressure in the boiler increases. If some form of safety device is not provided or is inadequate to limit the pressure to a safe value, the boiler will rupture. Boilers are required to be equipped with safety valves to relieve pressure if they exceed set values. Low points in some boilers are provided with fusible plugs. During normal operations, the plugs are covered with water which keeps them cool. If the water level drops and they are exposed, the plugs melt and create another vent for pressure relief. Boilers also have low-water shutoff devices which block in the burner fuel when the steam drum water drops below a certain level.

Pressure vessels do not have to be fired to be hazardous. Heat input can occur in other ways. The sun can heat outdoor pressure vessels, such as portable compressed gas cylinders and sample bomb cylinders, some that contain gases at pressures up to 2,000 psi at room temperature. These vessels should be stored or housed in shaded areas. For example, the vapor pressure of liquid carbon dioxide is 835 psi at 70°F and 2,530 psi at 140°F. Pressure vessels (cylinders) inside buildings should not be located near sources of heat, such as radiators or furnaces.

Dynamic Pressure Hazards

Dynamic pressure hazards are hazards caused by a substance in motion. The substance can be a fragment of metal from a ruptured vessel or a whipping hose. A possible source of injury is through a pressure-gauge failure. Sometimes the thin-walled Bourdon tube or bellows inside the gauge case fails under pressure due to metal fatigue or corrosion. Unless the gauge case is equipped with a blowout back, the face of the gauge will rupture first, hurling out pieces of glass and metal. A person standing in front of the gauge will be injured. A blowout back does not prevent failures but it ensures that no fragments will be propelled forward. Some boiler and furnaces buildings have blowout panels in case there is an explosion due to delayed ignition of unburned fuel gases. The blow out panel is a deliberate weak area of a building wall that releases pressure through that section of wall rather than allowing the whole building to explode.

Pressure release valves (PRVs), also called safety valves (see Figure 7-1), lessen the possibility of a rupture due to overpressurization. Flammable or toxic discharges from PRVs are directed to flare headers or vent lines where they constitute no danger. Some rail road tank cars have PRVs, depending on their cargo. A rail tank car filled with gasoline was left in the freight yard at Ardmore, Oklahoma, and remained there for two days. The second day was sunny and hot. The tank car was black and absorbed heat, the gasoline in the tank vaporized, creating pressure in the tank, and the safety valve released to the environment. Gasoline vapor escaped for several hours and blanketed a nearby neighborhood. Ignition occurred, then an explosion. The flame bounded along the ground in streaks and there were reports of clothing being ignited as far away as 350 feet from the tank car. All buildings within 400 feet were destroyed and buildings 1,200 feet away were damaged.

The pressures in full cylinders of compressed air, oxygen, or nitrogen are over 2,000 pounds per square inch gauge (psig). A cylinder weighs slightly more than 200 pounds. The force generated by gas flowing through the small opening created when the brass valve breaks off a cylinder can be 20 to 50 times greater than the cylinder weight. Cylinders with broken valves have taken off like rockets, reached a velocity of 50 feet per second in

Figure 7-1   Pressure Relief Valves on a Boiler

one-tenth of a second, and smashed through buildings and rows of vehicles. An Olympic sprinter can't out run it.

Whipping flexible hoses are dangerous. Worker's skulls have been crushed by the end fitting of a compressed-air line which was not properly tightened when the line was connected. The line separated under pressure and lashed about until it hit a worker in the head and crushed his skull. These types of accidents can occur with water hoses as well. A whipping line of any kind can break bones and damage equipment. All high-pressure lines and hoses should be restrained from possible whipping. Rigid lines should be preferred to flexible hoses, but if the latter must be used they should be kept as short as possible. If a line or hose gets free, workers should be trained to leave the area immediately and shut off flow to the line. They should never attempt to grab and restrain a whipping line or hose.

OTHER PRESSURE HAZARDS

Systems under pressure should not be worked on. Each pressurized vessel or line should be considered hazardous until all pressure has been released. Workers should verify a lack of pressure by checking a gauge directly connected to the vessel or line by opening a test cock. A person working on a nitrogen gas line pressurized to almost 6,000 psi failed to verify the line had been depressurized. He loosened the bolts of a flange. The bolts were those closest to him instead of those on the opposite side. The flange separated slightly and a thin stream of compressed gas at 6,000 psi shot out and cut into his leg like a knife. If a pressurized line is suspected of leaking, never use fingers or a booted foot to probe for the leak. High pressures can cut like knife blades. Safe alternatives are to use a piece of cloth on a stick (called a flag) or a soap-and-water solution.

Dirt, debris, and other particles can be blown by compressed gas into an eye and rupture it or through the skin like a bullet. Occupational Safety and Health Administration (OSHA) standards permit compressed air to be used for cleaning purposes if its pressure is less than 30 psi and if personal protective equipment is used.

There have been cases in which compressed air entered the circulatory system through cuts in the skin. Since the skin breathes through pores, compressed air can pass through the skin and into the blood stream. In a Massachusetts plant, a woodworker, covered with sawdust held a compressed air nozzle 12 inches away from the palm of his hand and opened the valve to blow the sawdust off. Within seconds, his hand swelled up to the size of a grapefruit.

Water Hammer

Water hammer (see Figure 7-2) is caused by a sudden stoppage of liquid flow. It is similar to a moving car hitting a brick wall. The sudden stop has a shock effect that creates a lot of pressure, which if severe enough, can cause the rupture of a line. The mass of water moving down a pipe has momentum. If the flow is terminated abruptly by closing a valve downstream, the momentum of the liquid is transformed into a shock wave (water hammer) which is transmitted back upstream. The shock is transmitted back through the liquid because liquids are practically incompressible. To avoid damage to liquid lines, avoid the use of quick-closing valves.

Transferred from en.wikipedia.org.

Figure 7-2   Waterhammer Damage to Steam Line Flex Coupling

Negative Pressure (Vacuums)

Unintended vacuums or negative pressures also can be dangerous and damaging. Many structures may not be built to withstand reversed stresses. Much of the damage done by high winds during hurricanes and tornadoes is due to negative pressures. Most buildings are designed to resist positive loads but not to resist negative pressures. Negative pressures might be generated on the lee side of a building when winds pass over it. Although the actual difference in pressure is very small, the area over which the total negative pressure will act is very large, so a large force is involved. For example, a roof on a small house may be 1,500 square feet. If the difference in pressure is only 0.05 psi, the force tending to tear off the roof equals 1,500 ft^2 x 144 in/ft^2 x 0.05 psi, or 10,800 pounds (about five tons).

Field storage tanks and rail car tanks have collapsed due to unintended vacuums. Condensation of vapors in closed vessels, such as a rail tank car, is a source of vacuum pressure which could collapse the vessels. A liquid occupies far less space than does the same weight of its vapor. A vapor that cools and liquefies, such as steam, will greatly decrease the pressure inside the vessel. Unless the vessel is designed to sustain the load imposed by the difference between the outside and inside pressures, or unless a vacuum breaker is provided, the vessel may collapse. Steam cleaning a rail car, then blocking in the steam and leaving for the day will collapse the car. The steam will condense to water leaving a vacuum. In the same way, draining an unvented storage tank may produce enough negative pressure to collapse the tank.

Detection of Pressure Hazards

Finding gas leaks (pressure hazards) can be difficult. After a gas has leaked into the ocean of air around it, obvious symptoms of the leak (odor or a cloud) may disappear. Several methods of detecting gas leaks are:

- Sound—the gas discharge may be indicated by a whistling noise, particularly with highly pressurized gases escaping through small openings. However, these sounds may be hard to hear in a noisy plant. Workers should be careful when searching for gas leaks as high-pressure gases can cut through clothing and flesh, even boots.
- Streamers (cloth or plastic banners) may be tied to a stick to help locate leaks.
- Soap solutions smeared over the vessel surface form bubbles when the gas escapes if the leak is small and of low velocity.
- Scents may be added to gases that do not naturally have an odor. This is what is done with natural gas before it is piped into homes.
- Portable leak detectors may be used.

HAZARDS OF STEAM

Steam is the utility most observed by people passing by refineries and chemical plants. White clouds of escaping steam are seen all over a plant because steam can be used for so many things. Steam:

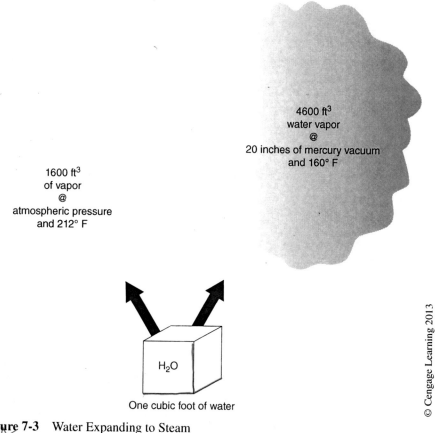

4600 ft^3
water vapor
@
20 inches of mercury vacuum
and 160° F

1600 ft^3
of vapor
@
atmospheric pressure
and 212° F

H_2O

One cubic foot of water

© Cengage Learning 2013

Figure 7-3 Water Expanding to Steam

- Furnishes power to turbines to drive pumps, compressors, etc.
- Supplies heat to reboilers, tanks, and kettles
- Is used as a purge gas for equipment
- Blankets vessels to prevent the formation of flammable mixtures
- Snuffs fires in furnaces and other equipment

All people who use microwave ovens or stoves realize that burns are an obvious hazard of steam. Burns are also a hazard of steam condensate. There are several other hazards. When water trapped in a process vessel is heated enough to become steam, the steam occupies 1,600 times the space of the water. Thus it can create a huge pressure wave (see Figure 7-3). Steam is not combustible and does not support combustion. For this reason, it is used as an inert gas. During unit shutdown, steam condensing on vessel surfaces helps to clean vessels and lines as well as to displace gaseous and liquid hydrocarbons before the equipment is opened. Steam is readily available in refineries and is usually cheaper than other inert gases, which makes it very desirable for a purging or blanketing medium. When steam enters a cold vessel, all or part of the steam condenses as water and drains to the lower part of vessel. Only after the equipment (metal) is thoroughly heated does the steam remain as an inert vapor. Because of condensation, however, steam introduced into a vessel may initially displace very little of the air or gas. A visible plume of steam at the purge vent is not a reliable sign that a vessel has been thoroughly purged of air. The temperature of a saturated steam-air mixture at any pressure is an indication of its air content. There are tables that can be consulted that indicate the temperature and corresponding volume percent air or other noncondensable in various mixtures at atmospheric pressure.

After vessels that operate at atmospheric pressure or above have been purged with steam prior to startup, fuel gas or another suitable gas must be backed (pumped) into the vessel when purging is completed. This is done to displace the steam, which if left inside, will condense and form a vacuum. If the vessel is left full of steam with valves closed, condensation can produce a vacuum great enough to collapse the vessel. Also, since valves frequently do not close tightly, the vacuum caused by condensing steam may draw in air. This creates a flammability hazard when hydrocarbons are introduced into the equipment.

Steam heating of blocked in exchangers or steam tracing of pipe or other equipment completely full of liquid can result in dangerously high pressures if a pressure relief valve is not provided. Liquids expand when heated. If a full vessel is blocked in and heated, something is going to give.

HAZARDS OF ELECTRICITY

Electrocutions kill an average of 143 construction workers each year. Data from 1992 through 2003 indicates electrical workers suffered the highest number of electrocutions per year, 34 percent of the total deaths caused by electrocution ("Alarming statistics," by D. Bremer, 2007, *Electrical Contractor*). Direct death from electrical shock results from ventricular fibrillation, paralysis of the respiratory center, or a combination of the two. Electricity can be hazardous to the process employees in a variety of forms. They are:

- Sparks and arcs
- Static electricity
- Lightning
- Stray currents
- Energized equipment

The two principle hazards of electricity are that it (1) might ignite mixtures of air and flammable gases or vapors, or (2) electrocute an employee or cause serious injuries.

Electrical Shock Injury

The amount of current and current path are two important factors affecting the extent of electrical shock injury. The amount of current depends on voltage and body resistance. Body resistance can be high or low, depending on whether the skin is dry (high resistance) or wet (low resistance). Contact area also affects body resistance: a person in a bathtub has both wet skin, a large contact area, and is almost certain to be electrocuted if a shock is received. Electrocution may occur when the heart area or the respiratory control center of the brain is in the current's path. The human body can only tolerate a very small amount of current and it is measured in milliamperes (a milliampere is one thousandth of an ampere). A rough guide published by the National Safety Council, is revealed in Table 7-1.

Table 7-1 Current and Injury

Current (milliamperes)	Injury
1 to 8	Shock sensation.
8 to 15	Painful shock.
15 to 20	Painful shock with control of adjacent muscles lost. Individual cannot let go.
20 to 50	Painful shock with severe muscular contractions and difficult breathing.
50 or more	May be fatal.

© Cengage Learning 2013

A few millivolts applied directly to the heart can cause fibrillation and death, yet it is a common (and unsafe) practice among some electricians to determine if a 120-volt or even a 240-volt circuit is energized by putting two fingers of the same hand in contact with the two conductors. If the person is insulated from the ground by standing on a dry wood floor, the current path is through the fingers and the hazard is negligible. If the person is standing in water or has a firm grasp on a metal water pipe, the current path will be through the central nervous system, which could be fatal. A lot depends also on the type of contact because an electric shock causes muscles to contract. A fingertip contact can be broken, but a grasped pipe might be impossible to release.

The significant factor is current flow through the body. A current of less than 1 milliamp (mA) may not even be noticed by a normal man. Above 3 mA, it becomes unpleasant. Above 10 ma, the victim is unable to let go. Above 30 mA, asphyxiation will result. Still higher levels lead to heart stoppage and death. These values are for sustained contact. Much higher levels can be tolerated for a fraction of a second.

Sparks and Arcs

Electric sparks and arcs occur in the normal operation of certain electrical equipment. They also occur during the breakdown of insulation on electrical equipment. When electricity jumps a gap in air, it is called a spark. We are all familiar with the static spark that jumps from the end of a finger to a metal doorknob after walking across a carpet. The minimum amount of energy which a spark or arc must have to ignite a flammable mixture is extremely small. Most electrical equipment can produce sparks and arcs which have more than enough energy to cause ignition.

Because of electrical inertia (inductance), an arc occurs when two contacts are separated. This inertia (inductance) simply means that electricity which is flowing tries to keep on flowing. Circuits which contain coils (such as electric motors) have a large amount of electrical inertia. The amount of current flowing when the contacts are first separated is also of great importance because current helps determine the intensity of the arc. Most large electrical switchgear equipment is contained in large substantial enclosures which are safety interlocked to prevent the door being opened while the breaker is in service. This is a safety feature to prevent accidental electrical contact and provide protection from arcing.

Working with 220/440 V switchgear can be dangerous because of the possibility of deadly arc flashes. Arc flashes occur when an arc shorts across components in a system and creates

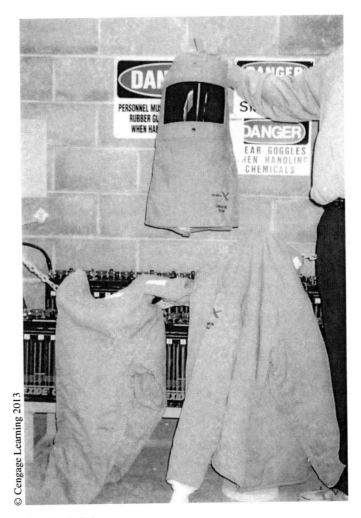

© Cengage Learning 2013

Figure 7-4   Arc Flash Suit

an ultraviolet flash that can permanently blind a person, plus temperatures of 35,000°F can occur, vaporizing metal into a gas, and heating the air so that it creates a pressure wave called an "arc blast." Bystanders have been killed standing eight feet away from the point of the arc flash. Special arc flash suits (see Figure 7-4) are available for operating switch-gear capable of an arc flash.

If an area is always hazardous due to the flammability of the chemicals in its processes, all the electrical equipment in that area must be enclosed in explosion-proof housings. Equipment which sparks or arcs during normal operation will sooner or later contact a flammable gas or vapor and the result will be a fire or explosion. To prevent this, electrical equipment which sparks or arcs during normal operation must always be explosion-proof even when used in an area which is only infrequently hazardous.

Static Electricity

The principal hazard of **static electricity** is a spark discharge which can ignite a flammable mixture. Refined flammable liquids, such as gasoline, kerosene, jet fuels, fuel oils, and similar products become charged with static electricity by the friction from pumping, flowing through pipes, splash a vessel, or by water settling through them. Different liquids generate different amounts of static electricity.

Refined hydrocarbon liquid fuels vary widely in their ability to generate and to conduct static electricity. Generally speaking, the products that are the better conductors also are better generators of static electricity, but because they are better conductors, the static electricity generated is discharged more readily. The discharge process is called *relaxation* and relaxation time is often expressed as the time required for a given charge to decrease to half its original value. If this time is very short, large static potentials in bulk fuel are not created because the relaxation process limits the charge that can build up. The poorer conductors (generally the cleaner products) are also poorer generators but because their relaxation times are so much longer, large charges can be generated. Very poor conductors, having excessive relaxation times, are also such poor generators that hazardous static potentials might not be created. Techniques used to reduce or eliminate static electricity as an ignition source are:

- Relaxation, a technique used when liquids from a pipe are discharged into the top of a vessel. The charge build-up can be reduced by enlarging the pipe diameter at the tip of the pipe, which reduces fluid velocity.
- Dip pipes reduce the static charge of non-conductive liquids in free-fall into a vessel. The pipe is extended close to the bottom of the tank, minimizing free-fall.
- Bonding or grounding

★Bonding and Grounding

Bonding and grounding are essential to electrical safety and used extensively in plants. **Bonding** means connecting two objects together with metal, usually a piece of copper wire. **Grounding** consists of connecting an object to the earth with metal, usually copper wire. The connection to earth is usually made to a ground rod or underground water piping. Electrical equipment is grounded first for protection of personnel, and second for the protection of equipment. In a refinery or petrochemical plant containing many large or tall metal vessels (flares, distillation towers, tanks, etc.) grounding is essential for protection from lightning.

Bonding allows a company to connect several vessels or pieces of equipment together and then to one ground rod. It is much cheaper than having a ground rod for each piece of equipment, plus, in some cases it may not be possible to drive a ground rod into the ground by the equipment. Grounding serves two distinct purposes relating to safety. First, since the ordinary power circuit has one side grounded, a fault that results in electrical contact to the grounded enclosure will pass enough current to blow a fuse. Second, the possibility of shock hazard is minimized since the low-resistance path of a properly bonded and grounded system will maintain all exposed surfaces at substantially ground potential. Grounding is effective against the hazard of leakage currents. Electricity, like water in a pipe, is always looking for a way out. Electricity is contained by insulation, but if the insulation is worn or frayed, electricity may leak out. All electrical insulation is subject to some electrical leakage which increases significantly as insulation deteriorates with age or as layers of conductive dust accumulate in the presence of high humidity. A proper grounding system with low electrical resistance will conduct leakage currents to ground without developing a significant potential on exposed surfaces.

Explosive Hazards of Electrical Equipment

An explosion is dependent upon the simultaneous presence of three conditions:

- Oxygen (air)
- Fuel, a gas vapor or fine solid (dust)
- Ignition source (arc, spark, heat)

Air is everywhere. In refineries and petrochemical plants, fuel is everywhere and contained in vessels and piping that has a potential to leak or rupture and create a hazardous atmosphere. Ignition sources, which are not as prevalent as fuel, are one thing management and workers can seek to control.

In processing and manufacturing sites that use flammable fluids, electrical equipment must be built and operated in a manner to prevent its becoming a source of ignition. It could ignite a hazardous atmosphere in either of two ways: by surface temperatures in excess of the ignition temperature or by sparks. Some sparks are part of normal operation, as in the operation of on/off switches. Some sparks are accidental, as in faulty connections. Both must be guarded against.

Explosion-proof electrical equipment (see Figure 7-5) presents no explosion hazard if properly installed and maintained. Such equipment is designed to withstand the pressure created by an internal explosion and to cool hot gases below ignition temperature before they reach the outside of the explosion-proof housing. This cooling is accomplished by routing

© Cengage Learning 2013

Figure 7-5 Explosion Proof Electric Panel Box

the escaping gas between closely machined flanges or threaded joints. An explosion-proof apparatus is not intended to be gas tight. It is a general assumption that no enclosure that has to be opened from time to time for inspection or maintenance can be maintained gas tight. This means that if the atmosphere surrounding it is hazardous, the atmosphere within will also be hazardous and result in an internal explosion. If the box withstands the explosion and the only openings are long, narrow, crooked paths, the escaping gases will not be hot enough to ignite the surrounding hazardous atmosphere. The primary danger results from carelessness by employees. Flange faces may become scratched or dirty, cover bolts may not be tightened or be missing or threaded covers may not be fully engaged. In such cases, the gases escaping from an internal explosion may still be hot enough to ignite flammable mixtures outside the explosion-proof housing.

Areas in which combustible gas, vapor, or dust may be present in explosive proportions are called *hazardous locations*. Special precautions must be taken with electrical equipment in hazardous locations to eliminate a source of ignition that could touch off an explosion. Hazardous areas are of two types:

1. Areas which are considered always hazardous because flammable gases or vapors will be present all or most of the time under normal conditions
2. Areas which are considered hazardous only infrequently as a result of ruptures, leaks, or other unusual circumstances

The National Electrical Code (NEC) classifies hazardous areas of various types and states what sort of equipment is safe for use in each. Refineries and petrochemical plants are most concerned with those areas which the NEC calls Class I. Class I locations are those in which flammable gases or vapors are or may be present in the air in amounts large enough to produce ignitable or explosive mixtures.

SUMMARY

The reduction of pressure hazards often requires better maintenance and inspection of equipment that measures or uses high-pressure gases. Proper storage of pressurized containers reduces many pressure hazards. Because the whipping action of pressurized flexible hoses can be dangerous, the hoses should be firmly clamped at the ends when pressurized. Pressure in vessels should be released before working on equipment and the vessels checked with gauges for signs of pressure. Negative pressures (vacuums) are pressures below atmospheric level. Vacuums can cause closed vessels to collapse unless the vessel has been designed to sustain negative pressures without harm.

Steam is not combustible and does not support combustion. For this reason, it is used as an inert gas. A great volume of steam may have to be used for purging because when steam enters a cold vessel or line, all or part of the steam condenses. Only after the equipment (metal) is thoroughly heated does the steam remain as an inert vapor. If a vessel has been steam purged and is left full of steam with valves closed, condensation of the steam can produce a vacuum great enough to collapse the vessel.

Electric sparks and arcs occur in the normal operation of certain electrical equipment. They also occur during the breakdown of insulation on electrical equipment. The minimum amount of energy which a spark or arc must have to ignite a flammable mixture is extremely small. Most electrical equipment can produce sparks and arcs which have more than enough energy to cause ignition. Bonding and grounding are essential to electrical safety and used extensively in plants to protect workers and equipment.

Electrical equipment must be built and operated in a manner to prevent it becoming a source of ignition. Explosion-proof electrical equipment is designed to withstand the pressure created by an internal explosion and to cool hot gases below ignition temperature before they reach the outside of the explosion-proof housing.

REVIEW QUESTIONS

1. Pressure is defined as:
   a. Force divided by area
   b. Force times pressure
   c. Force times area

2. Three types of hazards associated with high-pressure systems are _____, _____, and _____.

3. Three ways boilers are protected from over pressurization are by:
   a. Pressure release valves
   b. Fuses and breakers
   c. Fusible plugs
   d. Low water shutoff devices

4. Explain how a pressure gauge failure due to pressure can injure a worker.

5. (T/F) Pressure vessels have to fire to be hazardous.

6. The weakest point on a compressed air or nitrogen cylinder is the:
   a. Neck of the cylinder
   b. The bottom of the cylinder because of corrosion
   c. Cylinder valve

7. Explain how an operator should stop a whipping hose.

8. A _____ is a deliberate weak area of a building wall that releases pressure through that panel rather than allow a whole building to explode.

9. _____ is caused by a sudden stoppage of liquid flow that creates a lot of pressure.

10. (T/F) A whipping hose can break bones and damage equipment.

11. Condensation of vapors in a closed vessel is one method of creating a _____ in a vessel.

12. (T/F) Draining an unvented storage tank may collapse the tank.

13. Three ways to detect gas leaks are:
    a. Streamers
    b. Scanners
    c. Dosimeters
    d. Scents
    e. Soap solutions

14. A gallon of water converted to steam would occupy _____ times the space.

15. Explain why steam is commonly used as an inert gas for purging vessels and lines.

16. (T/F) Steam is combustible.

17. (T/F) A stream of high pressure gas can cut like a knife blade.

18. Explain one way to determine if steam purging has removed all air from a vessel.

19. Explain why steam purged vessels must be backfilled with fuel gas or any suitable gas.

20. Steam heating of a blocked in heat exchanger full of a liquid can create dangerously _____.

21. Three hazards of electricity are:
    a. Sparks and arcs
    b. Defective switches
    c. Stray currents
    d. Bernoulli flashes
    e. Static electricity

22. (T/F) A few millivolts applied directly to the heart can cause fibrillation and death.

23. List two dangers of an arc flash.

24. _____ reduce the static charge of non-conductive liquids in free-fall into a vessel.

25. _____ means connecting two objects together with metal.

26. _____ consists of connecting an object to the earth with a metal rod driven into the earth.

27. Describe how equipment is made explosion proof.

EXERCISES

1. Visit www.YouTube.com and find two videos on any subject in this chapter and write a one page report on what you learned from the videos. Cite your sources.

2. Search the Internet and find a PowerPoint presentation about water hammer, bonding and grounding, or static electricity. Download the presentation to a flash drive or CD and bring it to the class for presentation by yourself or your instructor. Cite your sources.

RESOURCES

www.wikipedia.org

www.osha.gov

www.YouTube.com

www.toolboxtopics.com

CHAPTER 8

Noise and
Vibration Hazards

Learning Objectives

Upon completion of this chapter the student should be able to:

- *State the fundamental hazard associated with excessive noise.*

- *List five factors that affect the risk of hearing loss.*

- *Explain the basic requirements for OSHA's Hearing Conservation Standard.*

- *Describe three types of hearing protection devices.*

- *Describe two effects of vibration on personnel or equipment.*

- *Describe five ways to reduce vibration in equipment.*

6 C P T A
↓
NAPTA

INTRODUCTION

The modern industrial worksite can be a noisy place. This poses two safety and health related problems. First, there is the problem of distraction. Any operation that requires oral communication will suffer from a noisy environment. Interference with communications can create misunderstandings about information transmitted from one person to another. When such communications relate to hazardous activities, any misunderstandings can lead to accidents. Second, there is the problem of hearing loss. Exposure to noise that exceeds prescribed levels can result in permanent hearing loss. Occupational exposure to loud sounds is the most common cause of what is often called *noise-induced hearing loss.*

Under the Occupational Safety and Health Administration (OSHA) Act, an employer who does not control noise so as to minimize fatigue and reduce the probability of accidents can be charged with a violation of the standard. Both employers and employees are obligated to observe existing noise standards. An employee who does not comply with previously described procedures for his or her welfare and suffers a loss of hearing can be charged with misconduct.

SOUND

Sound is any change in pressure that can be detected by the ear. Typically, sound is a change in air pressure. However, it can also be a change in water pressure or any pressure-sensitive medium. Noise is unwanted sound. Consequently, the difference between noise and sound is in the perception of the person hearing it. Loud music may be considered sound by a rock fan but noise by a shift worker trying sleep.

Sound and vibration are very similar. Sound relates to a sensation perceived by the inner ear as hearing. Vibration, on the other hand, is inaudible and is perceived through the sense of touch. The unit of measurement used for discussing the level of sound and what noise levels are hazardous is the decibel, or one-tenth of a *bel*. One decibel represents the smallest difference in the level of sound that can be perceived by the human ear. There are various weighting scales used to measure noise. We use the "A" weighting scale because it mimics the human ear. Noise standards are written for the dBA scale (decibel A scale). The OSHA Noise Standard requires monitoring instruments to be capable of measuring between 80 and 130 dBA. Table 8-1 shows the decibel levels for various common sounds. The weakest sound that can be heard by a healthy human ear in a quiet setting is known as the threshold of hearing (10 dBA). The maximum level of sound that can be perceived without experiencing pain is known as the threshold of pain (140 dBA).

Table 8-1   Decibel Levels for Common Sounds

Source	Decibels (dBA)
Normal conversation	60
OSHA level for required hearing protection	85*
Power saw	90
Chain saw	90
Passing truck	100
Compressor	100
Rock concert	110–120
Shotgun	140
Jet aircraft	150

© Cengage Learning 2013

*Eight-hour time weighted average

HEARING LOSS

Hearing loss is an impairment that interferes with the reception of sound and with the understanding of speech. The most important frequencies for speech understanding are those between 200 and 5,000 hertz. It is generally losses in this frequency range which are compensable under workers' compensation acts. A young person with normal hearing can detect sounds with a frequency range that extends from 20 to 20,000 hertz. Less than normal ability to hear speech indicates there has been degradation in the individual's hearing ability. Degradation of hearing can also result from aging, long-term exposure to sounds of even moderately high levels, or a sudden, very high intensity noise. The ear's greatest sensitivity is in the frequency range from 3,000 to 5,000 hertz, and the loss of hearing almost always occurs first at about 4,000 Hz.

Hearing losses in older persons were considered to be due to changes in the small bones of the middle ear, which caused a reduction in their ability to transmit higher-frequency vibrations. However, it is now believed that much of the loss formerly considered due to aging is actually due to almost constant exposure to loud sounds in our noisy modern society. This type of hearing loss involves deterioration of tiny ciliated cells in the inner ear. These cells convert the vibrations they receive to nervous impulses that are transmitted to the brain. When the ciliated cells in the inner ear are damaged by loud noises, the accompanying hearing loss is considered irreversible. Recent scientific studies have hinted at the possibility of the restoration of these cells. Occupational noise-induced hearing loss occurs over a period of several years' exposure to continuous or intermittent loud noise. Occupational noise-induced hearing loss almost always affects the hair cells in both inner ears, but occasionally the effect can be asymmetric. The loss is usually not profound, and once the exposure is removed, further hearing loss is prevented.

A very loud impulsive noise can cause ringing in the ears (tinnitus) and immediate loss of hearing sensitivity. If there is no further exposure to high noise levels, the tinnitus will disappear and hearing will return to the normal hearing level of the person exposed. Tinnitus also can be a result of the aging process. Tinnitus from aging might not disappear, and might occur almost continuously. In addition to hearing loss, excessive noise can cause physiological problems such as causing a quickened pulse, increased blood pressure, and constriction of blood vessels, all of which affect the heart and may lead to heart disease.

A number of different factors affect the risk of hearing loss associated with exposure to excessive noise. The most important of these are:

- Intensity of the noise
- Type of noise (wide band, narrow band, or impulse)
- Duration of daily exposure
- Total duration of exposure (number of years)
- Age

Of these various factors, the most critical are the sound level, frequency, duration, and distribution of noise. The unprotected human ear is at risk when exposed to sound levels exceeding 115 dBA. Exposure to sound levels below 80 dBA is generally considered safe. Prolonged exposure to noise levels higher than 80 dBA should be protected against through the use of appropriate personal protective devices.

To decrease the risk of hearing loss, exposure to noise should be limited to a maximum eight-hour time-weighted average of 90 dBA. Some general rules for dealing with noise in the workplace are:

- Exposures of less than 80 dBA may be considered safe for the purpose of risk assessment.
- A level of 90 dBA should be considered the maximum limit of continuous exposure over eight-hour days without protection.
- Continuous exposure to levels of 115 dBA and higher should not be allowed.
- Impulse noise should be limited to 140 dBA per eight-hour day for continuous exposure.

NOISE HAZARD STANDARDS AND REGULATIONS

The primary sources of standards and regulations relating to noise hazards are OSHA and the American National Standards Institute (ANSI). OSHA regulations require the implementation of hearing conservation programs (HCP) under certain conditions. OSHA regulations should be considered as minimum standards. ANSI's standard provides a way to determine the effectiveness of hearing conservation programs such as those required by OSHA.

In 1983, OSHA adopted a Hearing Conservation Amendment to 29 CFR 1910.95 that requires employers to implement hearing conservation programs in any work setting where employees are exposed to an eight-hour time-weighted average of 85 dBA and above. They are also required to provide personal protective devices for any employee who shows evidence of hearing loss, regardless of the noise level at his or her worksite.

Duration is another key factor in determining the safety of workplace noise. The regulation has a 50 percent 5 dBA logarithmic tradeoff. That is, for every 5 decibel increase in the noise level, the length of exposure must be reduced by 50 percent. For example, at 90 decibels, the sound level of a lawnmower, the limit of safe exposure is eight hours. At 95 dBA, the limit on exposure is four hours, and so on. For any sound that is 106 dBA and above (sandblaster or rock concert) exposure without protection should be less than one hour. Figure 8-1 shows the basic requirement of OSHA's Hearing Conservation Standard.

The following bullets explain the requirements of the hearing conservation standard:

- Monitoring noise levels—noise levels should be monitored on a regular basis. Whenever a new process is added, an existing process is altered, or new equipment is purchased, special monitoring should occur immediately.
- Medical surveillance—this component of the regulation specifies that employees who will be exposed to high noise levels be tested upon being hired and again at least annually.
- Noise controls—steps are to be taken to control noise at the source. Noise controls are required in situations where the noise level exceeds 90 dBA. Administrative controls are sufficient until noise levels exceed 100 dBA. Beyond 100 dBA, engineering controls must be used.
- Personal protective equipment (PPE)—specified as the next level of protection when administrative and engineering controls do not reduce noise hazards to acceptable levels. It is to be used in addition to administrative and engineering controls.

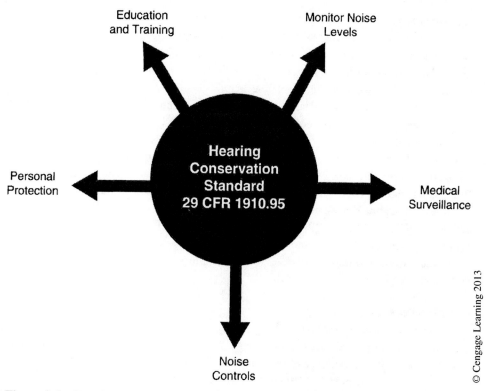

Figure 8-1   Requirements of Hearing Conservation Standard

- Education and training—this provision requires that employees understand (1) how the ear works, (2) how to interpret the results of audiometric tests, (3) how to select personal protective devices that will protect them against the types of noise hazards to which they are exposed, and (4) how to use personal protective devices properly.

NOISE ASSESSMENT

Assessing hazardous noise conditions in the workplace involve the following:

- Conducting periodic noise surveys
- Conducting periodic audiometric tests
- Record keeping
- Follow-up action

Noise Surveys

Conducting noise surveys involves measuring noise levels at different locations in the workplace. The devices that are most widely used to measure noise levels are sound level meters and dosimeters. A sound level meter produces an immediate reading that represents the noise level at a specific instant in time. A dosimeter provides a time-weighted average over a period of time such as one complete work shift. The dosimeter is the most widely used device because it measures total exposure, which is what OSHA and ANSI standards specify. Using a dosimeter in various work areas and attaching a personal dosimeter to one or more employees is the recommended approach to ensure dependable, accurate readings.

Audiometric Testing

Audiometric testing measures the hearing threshold of employees. Tests detect changes in the hearing threshold. A negative change represents hearing loss within a given frequency range. The audiogram from the test measures the noise threshold at which a subject responds to different test frequencies. These frequencies range from low to high tones. An audiometric examination can identify changes in hearing thresholds. Significant threshold shifts or loss of hearing might be attributed to overexposure to noise, either on or off the job. The purpose of conducting the noise survey is to identify employees who are exposed to high levels of noise and in danger of hearing loss. The initial audiogram establishes a baseline hearing threshold. After that, audiometric testing should occur at least annually. When even small changes in an employee's hearing threshold are identified, more frequent tests should be scheduled and conducted. Those employees found to have standard threshold shift—a loss of 10 dBA or more averaged at 2,000, 3,000, and 4,000 hertz (Hz) in either ear—the employer is required to make an entry in the OSHA 300 log in which the loss is recorded as a work time illness. Audiometric forms are completed and kept on file to allow for sequential comparisons.

Follow-Up after Audiometric Testing

A serious concern in occupational settings is failure to take appropriate action when the earliest stage of noise-induced hearing loss is observed. Because hearing loss can occur without producing any evidence of physiological damage, it is important to follow-up on even the slightest evidence of a change in an employee's hearing threshold. Follow-up can take any of the following forms:

- Administering a retest to verify the hearing loss
- Changing or improving the type of personal protection used
- Conducting a new noise survey in the employee's work area to determine if engineering controls are sufficient
- Testing other employees to determine if the hearing loss is isolated to the one employee in question or if other employees have been affected

Sometimes the hearing loss in both ears has increased significantly and the employee, for their protection, is moved to a process unit that doesn't have the potential for hearing loss.

Employee Training

OSHA standard 29 CFR 1910.95 requires that employers that have working conditions hazardous to their employee's hearing must train their employees. Training is required annually. Employees need to be trained to:

- Understand the danger to hearing that comes from noise exposure
- Recognize noise exposures which are harmful
- Evaluate noise levels of exposure in a practical way
- Take action to protect themselves from harm from noise

Many newly hired employees in processing and manufacturing industries do not realize how harmful noise is to their hearing. Because most hearing loss is gradual and takes place over a period of years, they don't notice the hearing loss until the loss is significant.

Annual training in the four categories mentioned previously and comfortable, easy to don hearing protection can prevent hearing loss.

HEARING PROTECTION DEVICES

When noise levels to which workers may be exposed exceed those indicated by the OSHA standards, personnel protection must be provided. However, long term use of hearing protection devices (HPDs) should not be resorted to until steps to reduce noise levels through engineering and administrative controls are exhausted. There are two basic types of HPDs: passive and active. The passive HPDs (see Figure 8-2) are the most common in industry and consist of several types of earplugs that are inserted into the ear canal and earmuffs that encircle the outer ear. Active HPDs consist of noise-attenuating helmets that provide active noise cancellation, communications features, and attenuation.

Plugs—are rubber or plastic devices that fit snugly against the ear canals, blocking the passages against transmission of sound. They come in a variety of sizes and types. Each person who must wear them should be fitted initially by a qualified audiologist, who will ensure they fit snugly and effectively and without discomfort. They are effective for eight-hour exposures only up to 95-dB noise levels. They provide no protection through the bony areas around the ears. Another disadvantage is that it is sometimes difficult for a supervisor to determine from a distance whether a worker is wearing them.

Foam plugs—are compressed and twisted between the thumb and forefinger, then inserted into the ear canal. The foam "memory" then causes the plug to try to return to its original shape, expanding it into the ear canal. The worker's skill in inserting these plugs

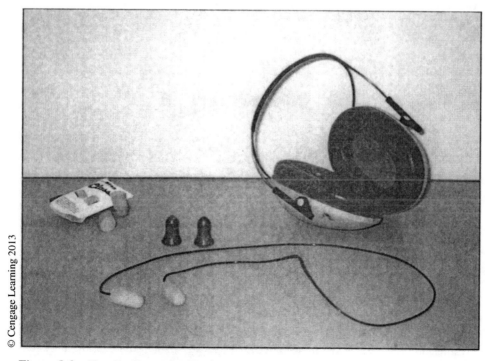

© Cengage Learning 2013

Figure 8-2 Hearing Protection Devices

properly is essential and requires employee training. This is one of the most common types of hearing protection.

Muffs—cover the entire ear and some of the bony areas around it through which sound might be conducted. They offer greater protection than do plugs which are not fitted properly. They are easily fitted and adjusted and easy to put on and take off. However, workers sometimes complain of headaches caused by the compression effect against the head, and they can be uncomfortable at high temperatures. Special types must be used if they are to be worn with hard hats or other headgear.

Active Hearing Protection Devices—are noise-attenuating helmets that provide active noise cancellation, communications features, and attenuation. They reduce noise by introducing destructive cancellation that applies opposite-phase sound waves at the ear. Some are designed for hearing protection and others for one- or two-way communication.

OTHER VIBRATION EFFECTS BESIDES NOISE

Noise is the most common vibration hazard; however, vibration can have other adverse effects. High-intensity, low-frequency sounds can cause the skull, other bones, and internal organs to become injured. Resonances will occur at certain frequencies so that these painful or injurious effects become much more noticeable. Vibration is transmitted more easily through solid materials than through air. It may happen, therefore, that a heavy piece of equipment can transmit vibrations through a structure, such as a frame and flooring of a building, to other equipment. Employees in contact with this equipment may become aware of and be affected by the transmitted vibration.

Even worse than the annoyance of vibration to personnel is the metal fatigue induced by vibration. Metal fatigue can cause failures of rotating parts and other stressed mechanical equipment. The result can be damage to equipment and possible injury to personnel. Vibrations can cause leakage of fluid lines, pressure vessels, and containers of hazardous liquids and gases. The use of vibrating tools can lead to arthritis, bursitis, injury to the soft tissues of the hands, and blockage of blood vessels. Of these, Raynaud's phenomenon, is probably the most prevalent and most serious. Raynaud's phenomenon involves paleness of the skin from oxygen deficiency due to reduction of blood flow caused by blood vessel and nerve spasms. Because of the deficiency in blood flow, the hands feel cold and may have decreased sensation. The disease is produced by vibration directly on the fingers or hands and has been associated pneumatic chisels and hammers and hand-held rotating grinding tools.

Engineers and designers should select equipment for installation which has low vibration and noise characteristics. They can require permissible maximum noise levels in specifications for new equipment. They can determine whether an operation, process, or piece of equipment that is noisy can be avoided or eliminated by use of a quieter one. Equipment that might vibrate should be mounted on firm, solid foundations. If equipment vibrates, they can determine whether or not its characteristics can be changed by use of devices such as dynamic dampers, rubber or plastic bumpers, flexible mountings and couplings, or resilient flooring. Where vibrations of fixed equipment cannot be eliminated, mount the equipment on vibration isolators to prevent transmission of motion.

SUMMARY

Noise is unwanted sound. It is a form of vibration conducted through solids, liquids, or gases. Noise can startle, annoy, and disrupt concentration, sleep, or relaxation. It can interrupt communication and interfere with job performance and safety, and it can lead to hearing loss and circulatory problems. Noise levels greater than 90 dBA should be avoided. Workers must wear hearing protection if workplace noise levels are greater than 90 dBA.

Several important factors that contribute to hearing loss are individual susceptibility, loudness of the sound, frequency of the sound, duration, and length of exposure. Limiting exposure, administrative and engineering controls, PPE, and employee training can prevent hearing loss. Employers are required to provide audiometric testing of employees, annual training on hazards of noise, establish a hearing baseline for each employee, and provide regular monitoring of employee hearing.

REVIEW QUESTIONS

1. Occupational exposure to loud sounds is the most common cause of _____.

2. The unit of measurement for the level of sound is the _____.

3. The maximum level of sound that a human can tolerate without experiencing pain is:
   a. 90 dBA
   b. 120 dBA
   c. 140 dBA

4. Describe how hearing loss occurs.

5. The maximum decibel level considered for continuous exposure over eight-hour days without protection is:
   a. 85 dBA
   b. 90 dBA
   c. 95 dBA

6. (T/F) Occupational noise-induced hearing loss occurs over a period of several years.

7. (Choose three) Exposure to excessive noise can lead to:
   a. Increased blood pressure
   b. Brain damage
   c. Constriction of blood vessels
   d. Heart disease

8. List the basic requirements for OSHA's Hearing Conservation Standard.

9. (T/F) Assessing hazardous noise conditions in the workplace involves conducting periodic audiometric tests and record keeping.

10. One device used to measure noise is the _____.

11. Explain how audiometric testing reveals information about an employee's ability to hear.

12. List three types of hearing protection devices.

13. List the four categories of employee training.

14. Hearing protection training is required every _____ years.

15. (T/F) Hearing loss can be detected over a short period of time.

16. Two effects of vibration on personnel are:
    a. Damage to internal organs
    b. Damage to hearing
    c. Arthritis and bursitis

EXERCISES

1. Go to www.YouTube.com and view two short videos on hearing protection and write a one page report on what you learned.

2. Search the Internet for a PowerPoint presentation on hearing protection, download it onto a flash drive or CD and bring it to class for you or your instructor to present to the class.

RESOURCES

www.wikipedia.org

www.osha.gov

www.YouTube.com

www.cdc.gov

www.cdc.gov/niosh

www.toolboxtopics.com

CHAPTER 9

Hazards of Temperature

Learning Objectives

Upon completion of this chapter the student should be able to:

- List five hazards of heat and cold.

- Describe the symptoms of heat stroke

- List five factors that make a person susceptible to heat stroke.

- Explain the first aid to be given to a person with heat stroke.

- Describe the symptoms of heat exhaustion.

- Explain the first aid to be given to a person suffering heat exhaustion.

- Describe three measures used to protect workers from cold hazards.

INTRODUCTION

Extremes of either heat or cold can be more than uncomfortable—they can be dangerous. Heat stress, cold stress, and burns are major concerns of in the processing industry. Employees who work outside during the summer on surfaces that heat up and store heat will get very hot. They will be around metal equipment and vessels that are hot and radiate large amounts of heat. The opportunity for heat stress will be present. Also, winter conditions on a processing unit in Colorado or Alaska will present the opportunity for frostbite or hypothermia. The prudent process employee will seek to understand the types of heat and cold stress and how to avoid them.

Heat is a form of energy indicated by temperature. Temperature extremes affect how well people work and how much work they can do. The human body is always producing heat and must remove excess heat in order to maintain body proper temperature. In the same way that excessive heat can affect process equipment and process safety, heat can also affect the human body and its proper functioning. During the summer, especially in geographic areas subject to hot summers, heat can become a serious safety hazard to a process technician.

By now you might be getting alarmed at the number of hazards that exist in processing and manufacturing work sites. You might be thinking that perhaps you don't want a job in the processing or manufacturing industries. Do not worry. For every hazard there is a method or methods of protection and we will soon be studying those chapters.

THE BODY'S RESPONSE TO HEAT

The human body maintains an appropriate balance between the metabolic heat that it produces and the environmental heat to which it is exposed. Sweating and the evaporation of the sweat are the body's primary way of trying to maintain an acceptable temperature balance. As sweat (water) evaporates it carries away latent heat, a large amount of heat removed when sweat goes from a liquid phase to a gas phase. However, when heat gain from the environment is more than the body can compensate for by sweating, the result is heat stress.

Sweating is the body's principal method for removing excess heat. Sweat consists of water and electrolytes (salts). An individual at rest and not under stress, sweats about one liter per day. The sweating rate for an individual under stress of heavy work or high temperatures is about four liters in four hours. The body must replace water and electrolytes to prevent heat stress or sickness. This is why many people who try to walk out of stranded situations in a desert with only a few liters of water fail and die. They do not realize how profusely they will sweat and how dehydrated they will become. In the pages that follow, we will take a closer look at temperature hazards the process technician may be exposed to in the processing industry.

The most common types of heat stress (see Figure 9-1) are heat stroke, heat exhaustion, heat cramps, heat rash, transient heat fatigue, and chronic heat fatigue. These various types of heat stress can initiate a number of undesirable bodily reactions, some very serious, including prickly heat, inadequate blood flow to vital body parts, circulatory shock, and cramps.

Heat Stroke

Heat stroke is a type of heat stress caused by a rapid rise in the body's core temperature. It is very dangerous and can be fatal. First aid should be rendered immediately. Heat stroke's symptoms are (1) hot, dry, mottled skin, (2) confusion and/or convulsions, and (3) loss of consciousness. In addition to these observable symptoms, the victim will have a rectal temperature of 104.5°F or higher.

Factors that make a person susceptible to heat stroke include the following:

- Obesity
- Poor physical condition
- Alcohol intake
- Cardiovascular disease
- Prolonged exertion in a hot environment

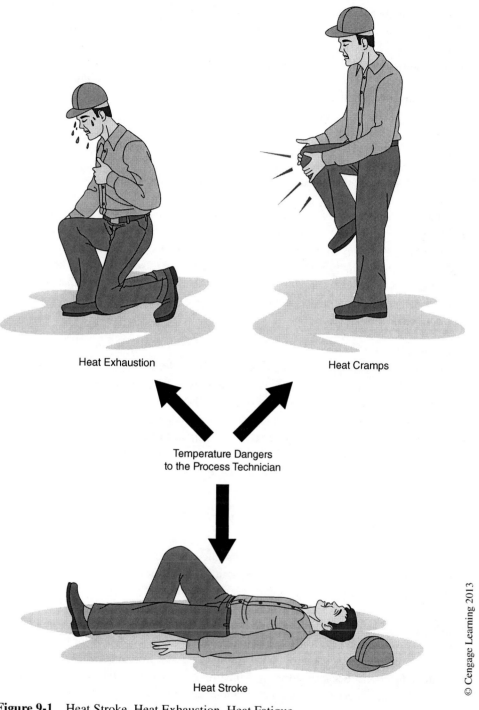

Heat Exhaustion

Heat Cramps

Temperature Dangers
to the Process Technician

Heat Stroke

© Cengage Learning 2013

Figure 9-1   Heat Stroke, Heat Exhaustion, Heat Fatigue

The last factor can cause heat stroke even in a healthy individual. A person who has one or more of the first four characteristics is even more susceptible to heat stroke. When heat stroke occurs, the body's ability to sweat becomes partially impaired or completely fails. Heat stroke can be fatal because sweating is the primary way the body disposes of excess heat. The inability to sweat causes the body temperature to increase uncontrollably. Action must be taken immediately to reduce a heat stroke victim's body core temperature. Do not wait for medical help. Immediately render first aid by immersing the victim in chilled water if it is available. If not, wrap the victim in a wet sheet and aim a fan set at high-speed at them. Add water periodically to keep the sheet wet.

Strategies to prevent heat stroke are:

- Gradual acclimatization to hot working conditions
- Monitoring and rotation of workers out of the hot environment at specified intervals during the work day
- Use of self-contained cooling personal protective clothing

Examples of special self-cooling clothing are (1) an ice vest that consists of a light fitting vest with 60 small pockets of ice, and (2) a vinyl one-piece coverall with a built-in air distribution system that directs cool air against the body and exhausts warm air. The ice vest is used when mobility is important and the length of exposure is short. The body suit is used when a longer exposure time is necessary. Also, many sites require that an outside worker come into an air-conditioned area on some scheduled frequency and also take salt tablets or some form of electrolyte.

Heat Exhaustion

Heat exhaustion is a type of heat stress caused by water and/or electrolyte depletion. A victim of heat exhaustion may have a normal or even lower-than-normal oral temperature but will typically have a higher-than-usual rectal temperature (i.e., 99.5°F to 101.3°F). Heat exhaustion is caused by prolonged exertion in a hot environment and a failure to replace the water and/or electrolytes lost through sweating. Heat exhaustion causes the body to become dehydrated, which decreases the volume of circulating blood. The various body parts compete for a smaller volume of blood, which causes circulatory strain. A victim of heat exhaustion should be moved to a cool place and made to lie down. Replacement fluid containing electrolytes should be taken slowly but steadily by mouth.

Prevention of heat exhaustion should be handled on the job in the same way as it is in professional sports. That is done by (1) gradual **acclimatization** to the weather, and (2) replacement of fluids and electrolytes lost by sweating.

Electrolyte imbalance is a problem with heat exhaustion and heat cramps. When people sweat, they lose salt and electrolytes. Electrolytes are minerals needed by the body to maintain its proper metabolism and for cells to produce energy. Loss of electrolytes interferes with these functions. Because of this, it is important to use commercially produced sports drinks that contain water, salt, sugar, potassium, or electrolytes to replace those lost through sweating.

Heat Cramps

Heat cramps are a form of heat stress caused by salt and potassium depletion. Observable symptoms are primarily muscle spasms that are typically felt in the arms, legs, and abdomen. Heat cramps are caused by salt and potassium depletion from heavy sweating due to working in a hot environment. Drinking just water worsens the problem because electrolytes are also necessary. During heat cramps, salt is lost and water taken in dilutes the body's electrolytes. Excess water enters the muscles and causes cramping. The victim should replenish their body's salt and potassium supply orally with commercially available drinks that contain carefully measured amounts of salts, potassium, electrolytes, and other elements. To prevent heat cramps, an employer should:

- Acclimate workers to the hot environment gradually
- Ensure that the fluid replacement product contains the appropriate amount of salt, potassium, and electrolytes

THERMAL BURNS

A serious hazard associated with heat in the workplace is burns. Burns can be very dangerous because they disrupt the normal functioning of the skin, which is the body's largest organ and the most important in terms of protecting other organs. To understand the hazards of burns you have to understand the function and composition of the skin. Human skin consists of the outer epidermis and the inner layer known as the dermis. The skin has several important purposes, some which are:

- Protection of body tissue
- Sensation
- Secretion
- Excretion
- Respiration

Protection from fluid loss, ultraviolet radiation, and infection by microorganisms is a major function of the skin, as are the sensory functions of touching, sensing heat and cold, and pain. The skin helps regulate body heat through sweating. What makes burns particularly dangerous is that they can disrupt any or all of these important functions, depending on the severity of the burn.

Severity of Burns

The severity of a burn depends on several factors, the most important being the depth of the burn. Did the burn penetrate the dermis or past the dermis into the subcutaneous layer? Other determining factors include location of the burn, age of the victim, and amount of burned area. Burns are classified by degree (i.e., first-, second-, or third-degree burns).

First-degree burns are minor and result only in a mild inflammation of the skin. Sunburn is an example of first-degree burn. The redness of the skin, the sensitivity of the skin, and moderate pain to the skin when touched make first-degree burns easy to recognize. **Second-degree burns** are recognizable from the blisters formed on the skin. If a second-degree burn is superficial, the skin will heal with little or no scarring. A deeper second-degree burn will form a thin layer of coagulated, dead cells that feels leathery to the touch. **Third-degree burns** are very dangerous and can be fatal depending on the amount of body surface affected. A third-degree burn penetrates through both the epidermis and the dermis. They can be caused by both moist and dry heat hazards. Moist hazards include steam and hot liquids that cause burns to appear white. Dry hazards include fire and hot objects or surfaces that cause burns to appear black and charred.

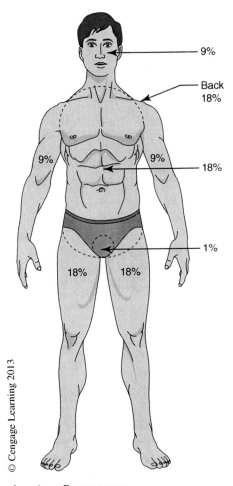

© Cengage Learning 2013

Figure 9-2 Body Surface Area Percentages

In addition to the depth of penetration of a burn the amount of surface area covered is also critical. The amount is expressed as a percentage of body surface area (BSA). Burns covering over 75 percent of BSA are usually fatal. Using the first-, second-, and third-degree burn classifications in conjunction with BSA percentages (see Figure 9-2), burns can be classified further as minor, moderate, or critical.

- All first-degree burns are considered **minor**. Second-degree burns covering less than 15 percent of the body are considered minor. Third-degree burns are considered minor if they cover only two percent or less of BSA.
- Second-degree burns that penetrate the epidermis and cover 15 percent or more of BSA are considered **moderate**. Second-degree burns that penetrate the dermis and cover from 15 to 30 percent of BSA are considered moderate. Third-degree burns can be considered moderate provided they cover less than 10 percent of BSA and are not on the hands, face, or feet.
- Second-degree burns covering more than 30 percent of BSA or third-degree burns covering over 10 percent of BSA are considered **critical**. Even small-area third-degree burns to the hands, face, or feet are considered critical because of the greater potential for infection to these areas. In addition, burns that are complicated by other injuries (fractures, soft tissue damage, etc.) are considered critical.

CHEMICAL BURNS

Many of the chemicals produced, handled, and stored by industry can cause first-, second-, and third-degree burns just like those caused by heat. All acids and bases can cause chemical burns, plus certain other chemicals. Chemical burns (see Figure 9-3), like thermal burns, destroy body tissues, however, chemical burns continue to destroy body tissue until the chemicals are washed away completely. Many concentrated chemicals have an affinity for water. When they contact body tissue (which is approximately 95 percent water) they withdraw water from it so rapidly that the original chemical composition of the tissue is destroyed. In fact, a strong caustic may dissolve even dehydrated animal tissue. The more concentrated the chemical solution, the more rapid the tissue destruction.

The severity of the burn produced by a chemical depends on the following factors:

- Corrosive capability of the chemical
- Concentration of the chemical
- Temperature of the chemical
- Duration of worker contact with the chemical

Effects of Chemical Burns

Different chemicals have different effects on the human body. The primary hazardous effects of chemical burns are infection, loss of body fluids, and shock. These effects are discussed in following paragraphs.

Infection—The risk of infection is high with chemical burns because the body's primary defense against infection-causing microorganisms (the skin) is damaged or destroyed.

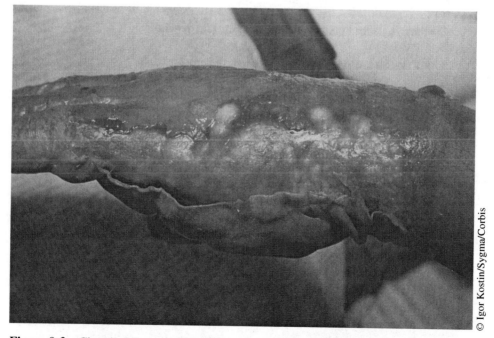

© Igor Kostin/Sygma/Corbis

Figure 9-3   Chemical Burn

This is why it is so important to keep burns clean. Infection in a burn wound can cause septicemia (blood poisoning).

Fluid Loss—Body fluid loss in second- and third-degree burns can be serious. With second-degree burns, the blisters that form on the skin often fill with fluid that seeps out of damaged tissue under the blister. With third-degree burns, fluids are lost internally and, as a result, can cause the same complications as a hemorrhage. If these fluids are not replaced the burns can be fatal.

Shock—Shock is a depression of the nervous system and can be caused by physical and/or psychological trauma. In cases of serious burns, the intense pain that occurs when skin is burned away and nerve endings are exposed may cause shock.

First Aid for Chemical Burns

In cases of chemical burns, the National Safety Council recommends washing off the chemical by flooding the burned areas with copious amounts of water as quickly as possible and for a minimum of 15 minutes. This is the only method for limiting the severity of the burn, and the loss of even a few seconds can be important. In the case of chemical burns to the eyes, continuous flooding (flushing) of the eyes should continue for at least 15 minutes. The eyelids should be held open to ensure that chemicals are not trapped under them. Another consideration when an employee comes in contact with a caustic chemical is their clothing. If chemicals have saturated the employee's clothes, they must be removed quickly. The best approach is to remove the clothes while flooding the body or the affected area with water. Most processing units are provided with safety showers and eye bubblers for protection against chemical burns.

HAZARDS OF COLD

The major injuries associated with cold conditions are either generalized (affects the whole body) or localized (affects a part of the body). A generalized injury from extremes of cold is hypothermia; localized injuries include frostbite and trenchfoot.

- **Frostbite** results from prolonged and severe vasoconstriction at temperatures below 32°F. In mild cases, tissue is not necessarily frozen. In more severe cases, ice penetrates the tissue and destroys it. Often, gangrene sets in if circulation has been severely reduced. The most vulnerable parts of the body are the nose, ears, toes, fingers, and cheeks. The first symptom of frostbite is a sensation of cold and numbness. Tingling and stinging sensations may follow this. Frostbite of the outer layer of skin results in a whitish, waxy look.
- **Trenchfoot**, also known as immersion foot, results from exposures below 53°F (12°C) for several days. Moisture from cold and sweat contributes to this hazard. Feet and legs become cold, pale, and numb, and cease to sweat. Nerve injury is frequent and loss of sensitivity may persist for weeks, even after the feet have been warmed.
- **Hypothermia** results when the body's core temperature drops to dangerously low levels. If the condition is not reversed, the person literally freezes to death. Some symptoms of hypothermia are uncontrolled shivering, sensation of cold, weakened pulse, slow or irregular heartbeat, and slow slurred speech.

In cold air, the body loses heat principally by radiation from exposed skin surfaces and a small amount of convection or conduction. The rate of heat loss increases with movement of air across the exposed skin, which produces a cooling effect. The heat lost from skin exposed to a 10-mile-per-hour wind when the ambient temperature is 10°F will be the same as that from skin in still air at -9°F. The wind chill factor is an indication of relative heat loss only. Freezing of tissue will not occur unless the temperature is 32°F or lower. For example, the wind chill factor for a 15-mile-per-hour wind at an ambient temperature of 40°F is 22°. Tissue and blood will not freeze, but the body will compensate for the added heat loss.

Some measures employers can use to protect workers from cold hazards are to install wind shields at some locations, provide heated shelters for warming breaks, provide warm drinks for fluid replacement, and rotate workers frequently.

SUMMARY

The most common forms of heat stress are heat stroke, heat exhaustion, heat cramps, and heat fatigue. Heat stroke is the most dangerous form of heat stress and can be fatal. The skin serves several important purposes, one of which is aiding in the regulation of body temperature. Burns can be especially dangerous because they disrupt the normal functioning of the skin, which is the body's largest organ and the most important in terms of protecting other organs. The severity of a burn depends on several factors, the most severe of which is the depth to which the burn penetrates. The most widely used method of classifying burns is by degree. Third-degree burns are the most critical.

Chemical burns, like thermal burns, destroy body tissues, however, chemical burns continue to destroy body tissue until the chemicals are washed away completely. The primary hazardous effects of chemical burns are infection, loss of body fluids, and shock.

The major injuries associated with cold conditions are either generalized or localized. A generalized injury from extremes of cold is hypothermia; localized injuries include frostbite and trenchfoot. Both frostbite and trenchfoot, if severe enough and left untreated, can lead to gangrene and amputation.

REVIEW QUESTIONS

1. (T/F) Sweating and evaporation is an important way for the body to regulate its temperature.

2. As sweat (water) evaporates it carries away_____, which removes large amounts of heat.

3. The body must replace _____ and _____ to prevent heat stress or sickness.

4. Three symptoms of heat stroke are:
    a. Hot, dry mottled skin
    b. Confusion and/or convulsions
    c. Loss of consciousness
    d. Profuse sweating

5. Four factors that make a person susceptible to heat stroke are _____, _____, _____, and _____.

6. Two types of first aid to be given to a person with heat stroke would consist of:
   a. Immerse the victim in chilled water, if possible
   b. Wrap the victim in a wet sheet and turn a fan on them
   c. Acclimate the victim

7. (T/F) Heat stroke can be fatal.

8. Describe the symptoms of heat exhaustion.

9. (T/F) Heat exhaustion is caused by prolonged exertion in a hot environment and a failure to replace water and/or electrolytes.

10. Explain what causes heat cramps.

11. Three important functions of the skin are _____, _____, and _____.

12. _____ are minor and result only in a mild inflammation of the skin.

13. Explain why burns can be very dangerous.

14. _____ penetrate through the epidermis and dermis.

15. _____ are recognizable from the blisters formed on the skin.

16. (T/F) Burns covering over 75 percent of the body surface area are usually fatal.

17. Chemical burns are like thermal burns because they _____.

18. The three primary hazardous effects of chemical burns are:
   a. Destruction of tissues
   b. Loss of body fluids
   c. Infection
   d. shock

19. Describe first aid for a chemical burn.

20. List three hazards of cold.

EXERCISES

1. Go to the Internet and research the various ways companies protect their workers from ambient high temperatures while working. Write a one page report on the ways and include a list of the sites visited.

2. Visit www.YouTube.com and view two videos on the hazards of heat stroke and/ or heat exhaustion. Write a one page report on the ways and include a list of the videos viewed.

RESOURCES

www.wikipedia.org

www.osha.gov

www.YouTube.com

www.toolboxtopics.com

CHAPTER 10

Hazards of
Process Sampling

Learning Objectives

Upon completion of this chapter the student should be able to:

- *Describe five physical and/or chemical hazards of samples.*

- *List three reasons why employees might have to use a gas detector.*

- *Explain why a vapor space is left in a sample container.*

- *Describe two atmospheres in which LEL meters should not be used.*

- *Explain how detector tubes reveal the presence and quantity of a specific gas.*

INTRODUCTION

This is a short chapter that briefly introduces you to the environmental, safety, and health hazards associated with the collection of samples, their transport, and ambient air sampling instrumentation (gas detectors). Process technicians and analyzer technicians will frequently be involved with collecting samples.

You might wonder, what is the big concern about catching a sample to warrant a whole chapter? If you work in the refining or petrochemical industry, your sample might contain components that might contain all the following hazards:

- Mutagen
- Teratogen

- Neurotoxic
- Flammable
- Toxic
- Carcinogenic
- Corrosive
- Allergen

Over the course of a year, some process technicians may collect hundreds or even thousands of samples. Over a 20- or 30-year career, a technician may collect tens of thousands of samples. Each sample collected, due to its chemical nature, may pose a health or physical hazard. Sample collection must always be done with care to avoid possible injury or health effects to the sample collector.

Besides collecting samples for process control reasons, process and analyzer technicians also use certain instrumentation to sample the air quality in certain areas, vessels, or during unit turnarounds. The instruments they use (gas detector) informs them of:

- Oxygen levels in a vessel (oxygen analyzer)
- Explosive levels of hydrocarbons in an area or vessel (combustible analyzer)
- The presence of toxic levels of gases in the air (detector tubes or analyzer)

A critical function of all process technicians is to be able to correctly operate and calibrate a portable gas detector.

HAZARDS OF COLLECTING SAMPLES

Our introduction mentioned that technicians collect many samples a year. Some of these samples may possess different health or physical hazards (see Figure 10-1). In fact, in a refinery or petrochemical plant it is fairly safe to say that most samples will have some type of hazard associated with them. So, why make a big issue about collecting samples? The answer is to promote the safety and health of the technician and ensure a non-toxic atmosphere in areas and vessels.

All samples must be collected in a safe and correct manner. Each sample must have a label denoting the sample ID, date, time, analysis required, and hazard warning. When you attached the sample labels to the sample containers, the labels have hazard warnings on them (mandated by the Occupational Safety and Healthy Administration' [OSHA] Hazard Communication Standard). They might say something like, "Warning: health hazard, benzene a known carcinogen." Or "Warning: flammable." Every sample collected has the potential to harm the collector, either in an acute or in a chronic way. If the technician was careless and did not wear their respirator when collecting a benzene sample, and did this several hundred times, over their long career, was it enough to initiate cancer (leukemia) after retirement? Or the technician expose their hands carelessly to just a drop or two of a neurotoxin (like phenol) when gloves weren't worn to collect the sample. Could that be enough to cause a nervous system dysfunction that forced them to walk with a walker shortly after they retired? This repeated exposure to very minute amounts of hazardous chemicals is called *micro-insults*. They cause no acute health effects, but they add up and eventually can cause chronic health effects. How many

120

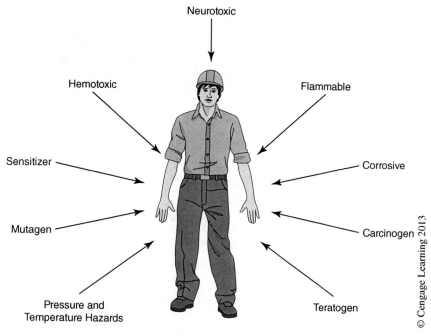

Figure 10-1 Sample Hazards

micro-insults can your body take before they are cumulative enough to cause serious harm? No one knows.

Micro-insult is a term that means exposure to small or supposedly insignificant amounts of a harmful chemical, such as one inhalation of a few parts per million of butadiene or one drop of benzene on the skin. Do this often enough and they add up. Management supplies you with the required personal protection to protect you from skin contact and inhalation. The law requires that technicians must use the personal protective equipment (PPE) provided in a responsible manner to prevent immediate and future injuries or illnesses.

Safety Precautions for Collecting Samples

A good technician always practices safe sample collecting. At some sites they are required to wear personal monitors or badges due to the hazardous nature of the chemicals of that process (see Figure 10-2). The following list includes some safety precautions that should be observed when collecting samples.

1. Use safety equipment (face shield, oxygen mask, glasses, gloves, etc.) whenever required.
2. Check the surrounding area for a source of ignition before sampling. Many flammable vapors are heavier than air and will move along the ground to an ignition source some distance away.
3. When sampling liquids, leave approximately a 20 percent vapor space above the liquid in the sample container to allow for thermal expansion.
4. Keep complete control of the sample flow when sampling.

© Cengage Learning 2013

Figure 10-2   Technician Wearing a Personal Monitor

5. Stand upwind when sampling so that any vapors that escape will be carried away from you.
6. Identify all samples with complete, readable tags or labels. Anyone who picks up a sample container should be able to identify the contents without difficulty.
7. Never use a leaking container or a container with faulty valves or other closures for sampling.
8. Place samples in a cool location as soon as possible.
9. Protect the valves on sample cylinders so that they are not accidentally damaged or unseated during transport to the laboratory. Do not use pliers or a wrench to tighten valves on sample cylinders. Excessive tightening will damage the valve and cause a leak.
10. Clean up any material spilled during sampling.

We close this topic with this statement. Collecting samples poses potential threats to the sample collector's health and safety, and to the environment. Do the task properly and responsibly.

GAS DETECTORS

Most process and analyzer technicians are trained to operate portable gas detectors. Sampling of the ambient air is an integral part of any safety, environmental, or health program for several reasons, some of which are to:

- Assure compliance to environmental regulations
- Detect fugitive emissions
- Protect employee and contractor safety and health

Many analyzers are housed in analyzer shelters and the sample is brought by tubing into the shelter to the analyzer. Analyzer shelters are like small houses and the sample lines are bringing the sample gases and liquids into the house to the analyzer. If any of the tubing lines or connections leak the shelter will accumulate hazardous vapors or gases. For this reason shelters have built in gas detectors (see Figure 10-3) that report to an alarm panel near the door of the shelter, warning the analyzer technician of the hazard of entry into the shelter.

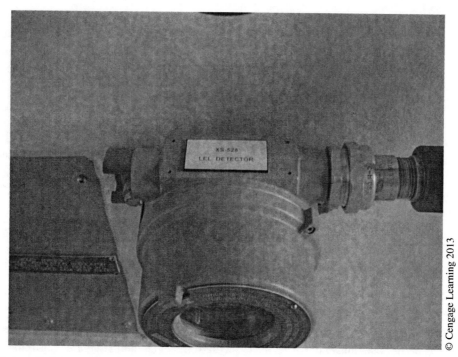

© Cengage Learning 2013

Figure 10-3   Analyzer Shelter lel/o2 Gas Detector

Process technicians can monitor the atmosphere of an area easily with several portable direct reading instruments, such as LEL/O_2 meters, volatile organic carbon (VOC) meters, and detector tubes. For personnel protection, personal exposure monitoring (wearing small badges or detectors) is considered the best choice. In addition, work area monitoring is also available. Area monitoring does not give a direct indication of personnel exposure, but gives information about unit or plant areas of potentially high exposures to certain gases, such as hydrogen sulfide or hydrogen cyanide.

Often, during unit turnarounds or maintenance of equipment, process technicians must gas test an area for combustibles before hot work can begin. Testing for the presence of gaseous or volatile hydrocarbons is performed:

- Routinely to detect fugitive emissions
- To prepare equipment or an area for hot work
- To prepare confined spaces for entry
- To prepare equipment for startup or shutdown

Combustible gas testing instruments indicate the volume percent of combustible gas in air expressed in percent of the lower explosive limit (% LEL). The LEL is the lowest percentage of the substance in air which will burn or explode when ignited. Most hydrocarbons will flash at lower hydrocarbon concentrations but will not generate enough heat to allow burning to continue.

Types of Gas Detectors

In recent years, a great variety of instruments have become available for air monitoring. The general trend in this type of instrumentation is miniaturization, ease of use, digital read out, and the capability of uploading the data to a computer. The simplification in detector calibration procedures has also favored the widespread use of the detectors. Many of the instruments are *active*, meaning they use a pump to draw a specified quantity of air into the instrument. Others are passive and use no pump. Generally, the accuracy of the active instruments is higher than that of the passive instruments. Almost all of the detectors are small, compact and portable, have solid-state electronics, and are very rugged. They have been designed for the rough conditions of the process industries.

Lower Explosion Limit (LEL) meters—also called combustible gas meters, are probably the most common direct reading instruments. In a broad sense, the LEL instruments are based on the heat released by the sample of gases when they are burned under controlled environment. Most instruments show the reading as the percent of the lower explosion limit (% LEL). LEL meters usually also contain an oxygen sensor and are commonly called LEL/ O_2 meters. There are several different designs but one of the most common is based on the heat of combustion of the burned hydrocarbon.

In the heat of combustion type there is a catalyst-coated filament that is kept heated at a certain temperature. When the gas sample passes over the filament, it is burned and the heat of combustion changes the filament resistance. This change in resistance also changes the electrical conductivity of the filament. This conductivity is proportional to the concentration of the contaminant in the air. The general practice for calibrating these instruments in industry is to calibrate them with specific gases (for example, methane) and use a table of response factors for other gases.

As a safety measure, analyzers with combustion sources have flashback arrestors to keep the flame from spreading out. Like many other instruments, LEL meters must be checked for zero reading before use. Like most other direct reading instruments, LEL meters are susceptible to steam or moisture. Most vendors advise not to use their LEL meters in steam-filled or highly humid atmospheres. Meters based on combustion need oxygen to function and would not work in an atmosphere devoid of oxygen. Many LEL meters have visual as well as audible alarms, digital displays, and self-diagnostics.

Oxygen meters—There are many models of oxygen meters available. Oxygen meters are calibrated on air (20.8 percent) which is usually marked on the meter scale as a calibration point. Often, LEL meters have an oxygen detection function built into them. Thus they can test for explosive atmospheres and for oxygen deficient atmospheres. For confined space operations, oxygen monitoring is a requirement.

Detector tubes—are probably, the simplest (and least accurate) of the direct reading instruments. The detector tubes are often referred to by the name of their manufacturer, such as Drager or Sensidyne tubes. The detector tube is hermetically sealed at both ends and filled with a granular or powdery reagent. Each tube is specific for a particular gas or vapor, hence tubes for ammonia detection are different from the tubes for carbon monoxide or methane detection. Each tube has printed on it's surface the name of the chemical it is specifically designed to test for.

Detector tubes are operated in conjunction with a hand pump or battery operated pump that draws ambient air into the tubes (see Figure 10-4). To sample the air, both ends of the tube are broken off in the special slots built into the pump body for that purpose, and the tube is inserted in the pump in the proper direction. An arrow on the tube indicates

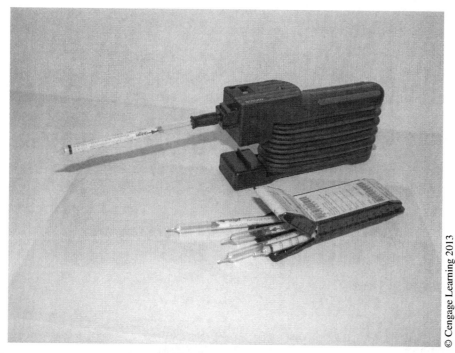

© Cengage Learning 2013

Figure 10-4   Drager Pump with Tubes

which end goes into the pump. A known amount of the ambient air sample is drawn into the pump. For hand pumps, directions will inform you how many times to pump the diaphragm; for battery driven pumps, there is a timer device.

The reagent in the tube (see Figure 10-5) is sensitive to the compound being tested for. The color of the reagent in the tube will change on contact with the compound. If the compound is present in the air sample, the tube develops a band of color and the length of the discolored reagent is proportional to the concentration of the contaminant. A tube is calibrated in either parts per million (PPM) or volume percent, levels that can be read directly from the tube where the colored band stops.

Detector tubes are convenient, low cost items but a technician should remember they are *estimation* devices. They cannot be used for some types of permitting applications, such as Confined Space Entry. Take the following precautions when using detector tubes.

- They are adversely affected by moisture or particulates. Though there is a guard section at the entrance to the tubes, moisture and/or particulate matter can quickly exceed the capacity of the guard section. Detector tubes should not be used in a rainy area, damp area, or dusty area.
- Before taking a reading from the tube, wait three to five minutes for the sample to diffuse completely through the tube.
- Follow the manufacturer's guidelines.
- Although the detector tubes are specific for specific gases, they are not immune to interference from other chemicals. For instance, tubes for benzene may be affected by the presence of phenol or toluene.

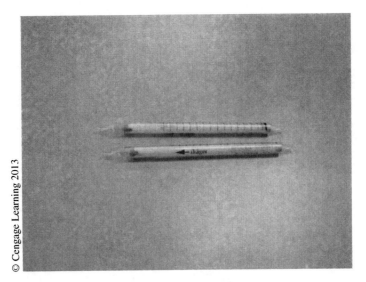

Figure 10-5 Drager Tube (close up)

SUMMARY

Management does not want to see their employees' hurt, sickened, or killed, if for no other reason than it is not good business. Workers tend to become complacent and careless about what they consider to be simple non-threatening tasks. Sample collecting should be considered a hazardous task, one a technician will be doing for 25 years. Outside of the physical hazards associated with some samples, there is the danger of micro-insults. Constant small exposures add up. A good technician always practices safe sample collecting.

One duty of a process technician is gas testing an area for combustibles before hot work can begin. Testing for the presence of gaseous hydrocarbons is performed routinely to detect fugitive emissions, to prepare confined spaces for entry, and to prepare equipment for startup or shutdown. Lower Explosion Limit (LEL) meters are probably the most common direct reading instruments. Most instruments show the reading as the percent of LEL. Often, LEL meters have an oxygen detection function built into them. Detector tubes are another type of gas specific detection device.

REVIEW QUESTIONS

1. List five physical and/or chemical hazards of samples.

2. (T/F) Over a 20–30 year career, a process technician may collect thousands of samples.

3. Three reasons technicians must use a gas detector on their process unit are:
   a. Check for combustible levels of hydrocarbons
   b. Check for oxygen levels
   c. Detect the presence of toxic levels of gases
   d. Prepare equipment or an area for hot work

4. Repeated exposure to minute quantities of hazardous chemicals that cause no acute health effects but may cause chronic health effects is called _____.

5. A _____ is left in a sample container as a safeguard against thermal expansion of the container.

6. An _____ is the lowest percentage of a substance in air which will burn or explode when ignited.

7. (T/F) _____ gas detectors have a pump to draw in atmospheric air.

8. Two atmospheres LEL meters should not be used in are:
   a. Highly humid
   b. Dust-filled
   c. Oxygen deficit

9. Describe how detector tubes reveal the presence and quantity of a specific gas.

10. (T/F) Detector tubes can be used for testing confined spaces.

EXERCISES

1. Visit www.YouTube.com and find two videos on fugitive emissions and write a one page report on what you learned. Cite your sources.

2. Research the Internet and locate a PowerPoint presentation about gas detectors and their use. Download the presentation to a flash drive or CD and bring it to class for presentation by yourself or your instructor. Cite your sources.

RESOURCES

www.wikipedia.org

www.osha.gov

www.YouTube.com

www.toolboxtopics.com

CHAPTER 11

Engineering Control of Hazards

Learning Objectives

Upon completion of this chapter the student should be able to:

- *List six types of engineering controls.*

- *Explain why engineering controls are the most desired way to eliminate hazards.*

- *Explain the hierarchy of hazard control.*

- *Explain how redundancy makes a system safer.*

- *Describe several types of process containment systems.*

- *Describe two types of process upset control systems.*

- *Explain how an explosion suppression system works.*

- *Describe two examples of isolation systems used in industry.*

- *Distinguish between a* safety interlock *system and a* process interlock *system*

INTRODUCTION

Chapters four through ten introduced us to the numerous health and physical hazards that may be present at a processing site. In this chapter and the next six, we reveal how industry seeks to eliminate or minimize these hazards. When assigned to a process unit and responsible for equipment on the unit, instruments or analyzers, the process employee must be trained to:

- Recognize the hazards in their area
- Determine whether the hazards can be reduced or eliminated
- Work safely with hazards that cannot be eliminated

This chapter familiarizes you with one important category of hazard and accident control, which is **engineering controls**. Engineering controls are the primary means of eliminating or minimizing hazards and include the general principles of enclosure, isolation, and ventilation. They consist of alarms, shutdown devices, and indication devices that alert the technician to a threatening situation. They help eliminate and/or minimize threats to safety, health, and the environment. Engineering controls are desirable because they are built into the process and are part of the process equipment. This reduces the need for human action to ensure safety.

HAZARD CONTROL

Hazard control involves the recognition, evaluation, and elimination or minimization of hazards in the workplace. Once hazards have been recognized and evaluated, safety and process engineers determine what can be done to protect process employees, the environment, and the community from these hazards. Engineers will determine if they can eliminate the hazard. However, sometimes the hazard cannot be eliminated due to expense or lack of technology. The expense may be too great for a corporation to bear. If the technology has not yet been invented, it means there is nothing that can be bought that will solve the problem. Management typically follows the hierarchy of hazard control listed below:

- Apply engineering controls, which are equipment-related methods of control
- Apply administrative controls, which are documented rules and procedures
- Require personal protective equipment (PPE)

Some of the questions asked in following this hierarchy of control are:

- Can a less toxic or non-toxic chemical be used instead of the present highly toxic chemical?
- If a less toxic chemical cannot be found, can the system be designed to greatly reduce or make nonexistent the risk of exposure to personnel?
- If exposure were still possible, what training and procedures need to be in place to further reduce the exposure?
- What PPE is required by the process employee to protect them from the hazard?

There is always some type of risk involved in a manufacturing facility. **Risk** is the possibility of loss or injury. Safety considerations are involved early in the design phase of a new plant to eliminate risk. Hazards are identified and plans designed to reduce or remove them. Emergency planning takes into account floods, hurricanes, inclement weather, evacuation routes, and the location of hazardous units. For example, in designing a new plant, the most hazardous process units would be located away from the most traveled and populated area of the plant.

Determination of Hazards

No job is 100 percent safe. Even secretaries get infected fingers from paper cuts. Each production process will have certain inherent hazards, however, the probability of accidents due to the hazards may be very small. Each process unit will have a limited number of primary hazards and a large number of initiating and contributory hazards. The two ways for determining the hazards associated with a task are (1) experience and (2) theory. Experience makes the hazards and safeguards for existing equipment and operations known. Theorizing can be used to estimate the hazards associated with proposed new products or modified operating procedures and equipment.

When an accident does occur, it often isn't simple to determine exactly which event was the root cause of the accident. What happened might be due to a complex series of events. For example, a high-pressure tank made of ordinary carbon steel ruptures violently, spewing hot flammable hydrocarbon into a work area and injures two operators. Moisture in the bottom of the tank caused corrosion, which reduced the strength of the metal, which ruptured. So, what was the root cause of the failure: moisture, corrosion, reduced strength, or pressure? In this series of events, moisture started the degradation process. If the tank had been made of stainless steel, there would have been no corrosion, moisture would not have been a problem, and there would have been no damage. Rupture of the tank, which caused the injury and damage, can be considered the primary hazard. Moisture started the series of events and can be called the initiating hazard. Corrosion, the loss of strength, and the pressure are contributory hazards. The primary hazard is often referred to by other names, such as a "catastrophic event" or "critical event." Some hazard terminology to become familiar with follows:

- Hazard—a condition with the potential to cause injury to personnel, damage to equipment, loss of material, or a lessening of the ability to perform a function.
- Primary hazard—a hazard that can directly and immediately cause injury or death, damage to equipment, disruption of plant operations, or accidental release of a large amount of chemical.
- Danger—relative exposure to a hazard.
- Safety—relative protection from exposure to hazards.
- Risk—expression of possible loss.
- Intrinsic safety—safety has been built into the design of the equipment.

Eliminating and Controlling Hazards

Hazards can be eliminated or controlled by good design or procedures for accident avoidance. The order of preference most commonly used for eliminating or controlling hazards is:

- Elimination of hazards is preferred.
- Where safeguards by manufacturer design are not feasible, protective safety devices should be employed (belt guards, railings, etc.).
- Automatic warning devices (signs, labels, flashing lights, etc.) should be added where neither design nor safety devices are practical.
- If none of the above is possible, clear concise procedures, PPE, and effective training become of primary importance.

TYPES OF ENGINEERING CONTROLS

There are many types of engineering controls (see Figure 11-1). Many are expensive and complicated but are worth the installation and maintenance costs because they prevent catastrophic failures that can result in the loss of valuable equipment and human life. The different types of engineering controls are:

- Fire alarms and detection systems
- Sprinkler and deluge systems
- Toxic gas alarms and detection systems
- Redundant alarms and shutdown devices
- Automatic shutdown devices
- Interlocks
- Process containment systems
- Process upset controls
- Ventilation
- Redesign of equipment to eliminate the hazard

Intrinsic Safety

The most effective method of avoiding accidents is through intrinsic safety. For a piece of equipment to be intrinsically safesafety has been built into the design of the equipment and is not an add-on function. Intrinsic safety can be achieved by (1) eliminating the hazard entirely or (2) limiting the hazard to a level below which it can do no harm. Under either condition, no possible accident can result from the hazard in question. Hazard elimination

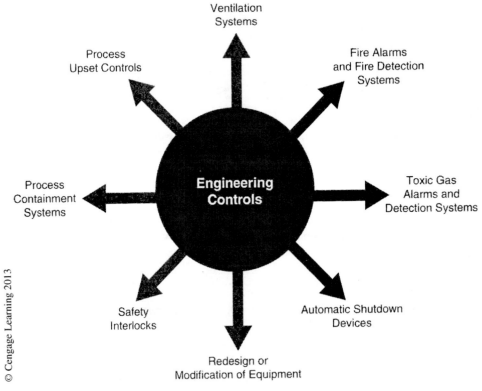

Figure 11-1 Types of Engineering Controls

means what it says, which is to create an environment free of hazards. Some examples of hazard elimination are:

- Good housekeeping, whether at home or at work, can prevent tripping over misplaced objects or slipping on wet surfaces by keeping facilities clean and orderly.
- Substituting pneumatic or hydraulic systems for electric systems where there is a possibility of fire or excessive heating.
- Rounding rough jagged or pointed edges and corners on equipment to prevent personnel injury.

If the hazard cannot be eliminated, then we can limit the level of the hazard so that no injury to workers or damage to equipment results. The energy available under any condition, even in a failure mode, should not be great enough to cause injury or death. Examples of methods to limit hazard levels include:

- Providing overflow arrangements to prevent excessively high liquid levels in storage tanks
- Ensuring the concentration of a flammable or toxic gas remains below a dangerous limit, and if the limit is exceeded, a blower starts automatically or in the case of a furnace with burner problems, detection devices (see Figures 11-2a and 11-2b) initiate a shutdown of fuel gas

© Cengage Learning 2013

Figure 11-2   (a) Flame Detector

© Cengage Learning 2013

Figure 11-2 (b) Gas Detector

- Incorporating pressure relief valves to keep pressure within a safe limit
- Using conductive coatings on materials to limit the level of accumulated static electricity

Fail-Safe Designs

Automatic shutdown devices (fail-safes) are activated by temperature, pressure, flow, level, and process composition values. They are an important part of any large system that is safety critical. Some fail-safe designs use a system, usually a microprocessor, that notices when process variable(s) are about to enter undesirable ranges and default into a fail-safe state. This puts the system in the safest possible state where it waits for the user to intervene or reset the system. Equipment failures produce a high percentage of accidents. Since failures are going to occur despite the best preventive maintenance programs, fail-safe arrangements are made to prevent:

- Injury
- Major catastrophes (fire, explosion, death, etc.)

- Damage to equipment
- Degradation of the process system

Fail-safe designs ensure that failures will leave the system unaffected or convert them to states in which no injury or damage will result. In most situations, this safeguard will result in the inactivation of the system. Fail-safe designs can be categorized into three types:

1. In the most common design, a *fail-passive* arrangement reduces the system to its lowest energy level. The system will not operate again until corrective action is taken. Circuit breakers and fuses for protection of electrical devices are examples of this type of fail-safe device. Solenoid valves (see Figure 11-3), such as this one on a steam control valve which is configured *fail close* shuts off instrument air, are another example.
2. A *fail-active* design maintains an energized condition that keeps the system in a safe operating mode until (a) corrective and overriding action occurs or (b) activates an alternative system to eliminate the possibility of an accident. A fail-active design might be a monitor system that activates a visual or audible indicator if a failure or adverse condition is detected in a critical operation. A battery-operated smoke detector is an example of a fail-active design.
3. A *fail-operational* design allows system functions to continue safely until corrective action is possible. This type of design is preferred since there is no loss of function. An example is the fail-safe operational orientation of the control rods on nuclear reactors, which automatically drop into place to reduce the reaction rate if it exceeds a preset limit.

© Cengage Learning 2013

Figure 11-3   Solenoid Valve on Steam Control Valve

Much railway equipment is based on fail-safe principles dedicated to the concept that gravity is a force that will always be there. As a result, semaphores, signal switches, and the lights to which they are connected are weighted devices. In the event of a power or system failure, a heavy arm is allowed to drop and activate the fail-safe warning signal. In the processing plants, if instrument air is lost control valves go to the more safe position—either fail open or fail closed.

Failure Minimization

If hazards are such that a fail-safe design is not feasible, steps to minimize the hazard are employed:

1. **Safety factors**—This is probably the oldest means to minimize accidents. Using this concept, components and structures are designed with strengths far greater than those normally required. This allows for calculation errors, variations in material strengths and stresses, material degradation, hurricanes, etc. If an item must withstand a prescribed stress, making it strong enough to withstand four times that stress would definitely reduce the number of failures. A structure with a safety factor of four would fail half as often as one with a factor of two.
2. **Failure-rate reduction**—This principle uses components and design arrangements to produce expected lifetimes far beyond the proposed periods of use. This reduces the probabilities of failure during operation.

Redundancy

Failure rates of complex equipment can often be greatly reduced through redundant arrangements. **Redundancy** provides multiple means of warning and shut down systems to ensure personnel and equipment are safe from failures that could result in a hazardous situation. Redundancy is based upon how critical an event is to worker safety or process continuation. Two separate switches or two separate alarms provide redundancy. In addition, there is standby redundancy. In this case, a piece of equipment operates until a failure is indicated, at which time another duplicate piece of equipment is turned on either automatically or manually. An excellent example of redundancy is fire water pumps. There are always at least two or more. A unit does not want to experience a fire, the site has only one pump, and that is the day the pump fails.

Weak Links

A **weak link** is one designed to fail at a level of stress that will minimize and control any possibility of a more serious failure or accident. Weak links have been used in electrical, mechanical, and structural systems for safety purposes for decades. The most common use of a weak link has been with electrical fuses that have been designed to fail before more valuable equipment is damaged. During a short circuit, the heat generated by passage of current through a metal with a low melting point causes the circuit to open before the current load becomes dangerous. Examples of weak links used outside of the processing industry are shear pins in outboard motors and brush hogs.

Other means of limiting extensive damage include:

● Boilers with mechanical fuses that melt when water levels drop excessively and allow steam to escape, thus preventing a rupture

136

Figure 11-4 Burst Rupture Disk

- Sprinkler systems that open to release water for fire extinguishing
- Shear pins that fail at designed stresses to prevent damage to equipment
- Rupture disks that release excessive pressure to the flare system (see Figure 11-4)

The problem with weak links is that they make the system inoperable and each activated weak link must be replaced before the system can start up again. Because of this, in some extremely critical designs weak links are used as secondary protection.

Alarms

Alarms are devices that alert the process technician to an abnormal or out of specification condition. They are usually audible (buzzer or horn) and visual (flashing light or flashing symbol or color on a computer screen). All interlocks should have an alarm to alert the process technician to a serious problem. Ideally, alarms associated with a safety interlock would be on a separate control panel.

Process Containment Systems

The function of process containment systems is to contain liquids or vapors and prevent them from entering the environment or creating a hazard. Examples of the different types of containment systems are:

- **Closed systems**—have vent/drain systems that prevent the release of vapors, gases, and liquids into the atmosphere.
- **Closed loop sampling**—benefits waste reduction and the environment because the sample stream is returned to the process via piping, reducing or eliminating any release of waste into the atmosphere.

- **Floating roof tanks**—the roof floats on the surface of the liquid and the seal between the roof and the sides prevents vapors from being released to the atmosphere. A floating roof tank may also have an external fixed roof to help eliminate toxic vapors when the vapor space is vented to the atmosphere. The vapor space between the external fixed roof and the floating roof contains a nitrogen blanket so that when the tank does vent only nitrogen is released to the atmosphere.

- **Ventilation systems**—consist of air handling and vent hoods, air movers, and positive pressure environments. Ventilation is used for control of airborne contaminants. Both local exhaust ventilation and general exhaust ventilation are used. Local ventilation is more desirable because it follows the rule that the control should be as near to the source as possible. As an example, a worker may use a grinder that produces fine metals particles and fumes that are harmful to inhale. The worker could operate the grinder under a ventilation hood (general ventilation) but there would still be the possibility of some exposure. Having a local exhaust outlet on the grinder would be the preferred method.

- **Safeguards against electrical shock**—are static dissipation devices, ground-fault circuit interrupters, and grounding and bonding mechanisms.

Process Upset Control Systems

When a process becomes upset, not just off-specification, but in danger of a fire, explosion, or chemical release, there are process control systems designed to handle the upset. Examples are:

- **Flares** (see Figure 11-5)—are basically a safety relief system used to burn waste gases and for the emergency dumping of gases from an upset unit. Toxic and flammable gases cannot be vented to the atmosphere because of safety, health, and environmental concerns. Vented gases empty into a flare header and are swept into the flare and burned.

- **Pressure Relief Valves (PRV)**—protect vessels or piping from over pressuring. They open when excessive pressure threatens a vessel. The material released is normally vented to a line that empties into the flare header.

- **Deluge systems**—are similar to very large sprinkler systems and are used to extinguish fires and suppress toxic releases. Deluge systems are expensive and are used primarily on processes concerned with extremely hazardous chemicals, such hydrofluoric acid. Most are activated automatically.

- **Explosion suppression systems**—usually are walls designed to contain explosions. The walls are called blast walls and may be built around a particular unit or vessel that contains very unstable materials or reactive chemicals. Also, if the vessel contains very unstable or reactive chemicals the vessel may be situated in cement-lined pit. Thus, the cement walls and earth act as a blast shield.

- **Dikes and curbs**—are simple devices that contain or direct spills. Dikes encircle storage tanks and should be able to contain the entire contents of the ruptured or leaking tank. Curbs are essentially mini-dikes that contain and/or direct small spills to sewer.

© Cengage Learning 2013

Figure 11-5 Industrial Flare

MONITORING DEVICES

Monitoring devices, quite simply, keep an eye on things. They can be used to keep any selected parameter, such as temperature or pressure, under surveillance to ensure it remains at proper levels, does not reach dangerous levels and no contingency or emergency is imminent. Monitors provide large benefits if problems are prevented or corrected immediately. Once a site has an emergency event, it incurs large expenses through responding to the emergency. Monitors can indicate whether:

- A specific condition does or does not exist
- The system is ready for operation and is operating as programmed
- The measured parameter is normal or abnormal
- A desired or undesired output is being created
- A specified limit is being met

Besides alerting personnel to a specified condition, a monitoring system must lead to suitable corrective action when necessary. This may be as simple as alerting an operator who is required to respond to the alarm (see Figure 11-6).

© Cengage Learning 2013

Figure 11-6   Control Room Alarm Panel

© Cengage Learning 2013

Figure 11-7   Fixed Sensor with Flashing Lights and Buzzer (Courtesy of Sensidyne Corporation)

Another type of monitor is a field mounted hazardous or toxic gas detection device. Such permanent gas detection systems (see Figure 11-7) protects personnel and equipment from virtually any toxic or combustible gases through audible and visual alarms in the field and/or on a control board. Permanent gas detection systems consist of a combination of transmitters, intelligent sensors, controllers, and accessories. These types of monitors are used to:

- Determine the presence of toxic gases, such as hydrogen sulfide and hydrogen cyanide
- Detect airborne levels of flammable substances
- Indicate with infrared detectors the presence of hot spots or flames
- Determine emissions of pollutants from stacks

ISOLATION, LOCKOUTS, LOCKINS, AND INTERLOCKS

Isolation, lockouts, lockins, and interlocks are some of the most commonly used safety measures in the process industry. These controls are constructed on three basic principles, or else on combinations of the first two. The three principles are:

1. Isolating a hazard once it has been recognized.
2. Preventing incompatible events from (a) occurring, (b) from occurring at the wrong time, or (c) from occurring in the wrong sequence.
3. Providing a release after suitable and correct action has been taken.

Isolation

Isolation employs separation as an accident prevention measure. Incompatible conditions or materials that would create a hazard if brought together are kept separated. Examples of isolation are:

- The separation of oxidizers from flammables
- The separation of stored acids and bases
- Isolating workers inside protective clothing
- Using explosion-proof electrical equipment in flammable atmospheres

An operational activity that generates a problem can be isolated from personnel who can be injured or equipment that can be damaged. A compressor producing a great amount of vibration and noise can be isolated through the use of vibration mounts and noise suppressors. Machine guards are widely used to isolate physical hazards in industrial plants and security fences around electrical substations serve the same function. Isolation can also be used to minimize the results of a violent release of energy. For instance, in production facilities that make or store explosives, distance is used as an isolating device. Possible points of catastrophic accidents are located far from persons, equipment, or vulnerable structures.

Lockouts and Lockins

Lockouts and lockins include some of the most common means of providing isolation of personnel, equipment, and operations in the process industry. Lockouts and lockins range from extremely simple devices such as bars on doors to bioidentification keypads in high security areas. The difference between lockouts and lockins is relative. A lockout prevents an event from occurring or prevents a person or other factor from entering an undesired zone. A lockin keeps a person or other factor from leaving a restricted zone. Some examples of lockouts and lockins are:

- Locking a switch on a circuit to prevent it from being energized is a lockout.
- Locking a switch on a live circuit to prevent current from being shut off is a lockin.
- Car sealing a valve closed to prevent fluid flow through it is a lockout.

Interlocks

Interlocks are commonly used safety devices. The function of an interlock is to prevent the occurrence of an event in the presence of certain conditions. Some interlocks prevent action or motion, others send signals to other devices that prevent the action or motion. They automatically reconfigure or interrupt final control devices if monitored variables deviate significantly from specifications. Typical process variables monitored are flow, pressure, level, and temperature. Typical machine variables monitored are coolant level and temperature, lubricant level and temperature, vibration, speed, etc. Interlocks allow equipment to start and operate only when monitored variables are within designed specifications. Interlocks inhibit unanticipated actuation of equipment and ensure correct startup/shutdown sequences are followed. A **permissive interlock** will not allow a process or equipment to startup unless certain conditions are met. There are two types of interlocks—safety and process interlocks. Each serves a different function.

Process interlocks make up an automatic system that detects an abnormal condition and either halts process action or takes corrective action to return the process to normal. Process interlock systems may be part of the Basic Process Control System (BPCS). The BPCS is a system of measurements and controls including alarms and interlocks that function to keep the process within acceptable operating limits. The BPCS is usually associated with producing good quality or in-spec finished product.

A **safety interlock** control function must be separate from the BPCS. Its function is not on-spec product but the prevention of a catastrophic event that would result in human injury or death or damage to equipment. Safety interlocks are usually hardwired to make it difficult to bypass or defeat them. This is done because there have been past occurrences of a unit engineer or a process technician jumping to the conclusion that an alarm was faulty and there was no problem. They tried to go around the interlock to shut off the alarm or prevent process interference. Safety interlocks must not be bypassed without written approval.

Depending on the degree of potential catastrophe, there usually is more than one safety interlock for a potential catastrophic event. Each of these safety interlocks including the sensor/transmitter, control function, and final control element are usually independent of the other safety interlocks for the same event. For maximum protection, each sensor for the same event should be unique to eliminate the potential of a common failure. The safety interlock must be fail-safe. This means that any loss of interlock power—electricity, air, hydraulics, etc.—loss of signal, must produce the same action as the safety interlock produces when it is activated (tripped).

Because a process technician or engineer cannot tell the difference from a process interlock and a safety interlock, there must be a list available to employees of the interlocks and their type. This list is usually part of the process safety information required by OSHA. Safety interlocks must be tested and the test documented. Often, this is on a yearly basis and done by an outside contractor.

SUMMARY

Engineering controls are the primary means of eliminating or minimizing hazards and include the general principles of enclosure, isolation, and ventilation. Engineering controls consist of alarms, shutdown devices, and indication devices that alert us to the buildup of a threatening situation. They help eliminate and/or minimize threats to safety, health, and the environment. Engineering controls are desirable because they are built into the process and can be made part of the process equipment. This reduces the need for human action to ensure safety. The typical hierarchy of hazard control begins with applying engineering controls where possible. If they are not possible or feasible, then administrative controls are applied. The last and least desirable control of a hazard is by requiring personal protective equipment.

Some of the more common engineering controls are fire alarms and detection systems, toxic gas alarms and detection systems, redundant alarms and shutdown devices, automatic shutdown devices, and process containment systems.

REVIEW QUESTIONS

1. _____ help eliminate or minimize threats to safety, health, and the environment.

2. List four types of engineering controls.

3. Explain why engineering controls are the most desired way to eliminate hazards.

4. (T/F) Engineering controls reduce the need for human action to ensure safety.

5. (Number the choices below in the correct order) The hierarchy of hazard control is:
   _____ administrative controls
   _____ engineering controls
   _____ PPE

6. _____ is the possibility of loss or injury.

7. A _____ is a condition with the potential to cause injury to personnel, damage to equipment, or loss of material.

8. For a piece of equipment to be _____ means safety has been built into the design of the equipment.

9. (T/F) Equipment failures produce a high percentage of accidents.

10. An example of a fail-safe design that is *fail-passive* is a _____ .

11. _____ provides multiple means of warning and shutdown systems to ensure personnel and equipment are safe from failures that would result in a hazardous situation.

12. _____ are devices that alert employees to an abnormal or out of specification condition and usually alert audibly and visually.

13. The function of _____ is to contain liquids or vapors and prevent them entering the environment or creating a hazard.

14. List three examples of process upset control systems.

15. _____ employs separation as an accident prevention measure.

16. A _____ will not allow a process or equipment to startup unless certain conditions are met.

17. An _____ reconfigures or interrupts final control devices to prevent a catastrophic event that would result in human injury or death or damage to equipment.

EXERCISES

1. Make a list of ten or more engineering controls you encounter in everyday life (at home, lawn equipment, automobiles, office buildings, work, etc.) and explain the function of each.

2. Research on the Internet one of the two subjects: a (1) deluge system or a (2) blast wall. Write a one page report.

RESOURCES

www.wikipedia.org

www.osha.gov

www.YouTube.com

www.toolboxtopics.com

CHAPTER 12

Administrative Control of Hazards

Learning Objectives

Upon completion of this chapter the student should be able to:

- *Define administrative controls and their purpose.*

- *Describe six types of administrative controls.*

- *List five administrative control activities.*

- *Explain the safety aspect of good housekeeping.*

- *Explain how audits differ from inspections.*

- *Describe the purpose of industrial hygiene monitoring.*

INTRODUCTION

Earlier chapters introduced us to the many types of hazards in the workplace. Controlling those hazards involves the recognition, evaluation, and elimination of the hazard in the workplace. If the hazard cannot be eliminated, it must be minimized as much as possible. When hazards cannot be engineered out of the workplace, **administrative controls** are the next logical step in hazard reduction. Administrative controls are procedures put into place to limit employee exposure to hazards. Their function is to influence worker behavior to follow operational, safety, and environmental procedures. Enforcement of the procedures may have penalties that vary in severity but can include employee termination of employment.

Administrative controls (see Figure 12-1) can be broken down into two broad categories: **programs** and **activities.** Programs are the written documents that explain how hazards

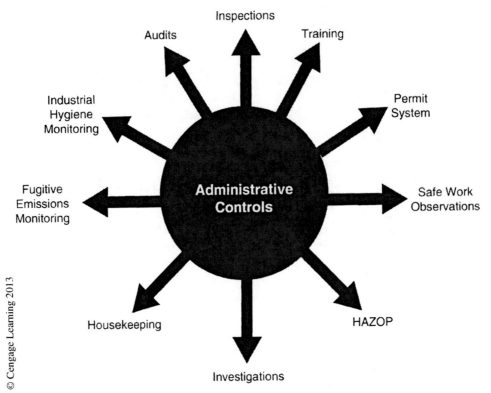

Figure 12-1 Types of Administrative Controls

are to be controlled. Activities are the programs put into action. Some examples of administrative controls and their categories are:

- Training—program/activity
- Housekeeping—activity
- Permit System—program
- Inspections (safety)—activity
- Audits (safety)—activity
- Investigations (safety)—activity
- HAZOP (Hazards and Operability study)—program/activity
- Industrial Hygiene monitoring—activity
- Fugitive emissions monitoring—activity
- Safe work observations—activity

The responsibilities of employees, contractors, and supervisors are clearly defined in the administrative procedures. Some procedures, rather than identify individual responsibilities, state "all employees and contractors are required to follow this procedure." Employees are trained on administrative procedures when hired and receive refresher training on a determined periodic basis. Hard copies or electronic copies of the procedures should be readily accessible to employees.

ADMINISTRATIVE PROGRAMS

Administrative programs are the written documents that explain how hazards are to be controlled. Many of them you see framed in hallways of hotels and displayed in waiting rooms of business offices (building evacuation plans). New hires, during orientation and training. are educated on site policies and standard operating. A partial list of the more common administrative hazard control programs includes:

- Policies, which in effect, are a business's guiding principles
- Procedures, which are step-by-step instructions for accomplishing a task
- Plans, which are methods prepared in advance for carrying out actions
- Principles, which are sets of rules or standards
- Rules, which are statements of how to do something or what may or may not be done

If a particular corporation has ten manufacturing sites, each site may have varying administrative programs, depending on plant size, equipment, and what is being manufactured. Exactly what is included in a written program depends on such varied things, as:

- Regulatory requirements (state or federal)
- Company-specific requirements
- Site-specific or plant-specific requirements
- Unit-specific or department-specific requirements

There are many written programs common to almost all process industries. One of the most common is the Hazard Communication Program, necessary in any facility where workers handle chemicals or are exposed to chemicals. Many administrative programs require employers and employees to perform certain activities in order to fulfill the intent of the program. In other words, there is a performance aspect (training) to the program besides a knowledge aspect. Workers must demonstrate that they not only have the required knowledge but also know how to use the equipment that will control the hazard, such as how to don a respirator.

Evacuation and Accountability Plans

These are established procedures that outline evacuation routes and assembly areas for prompt evacuation and accounting of onsite personnel in the event of an emergency. In many buildings open to the public, such as hotels, we see evacuation routes posted on hallway walls. Manufacturing plants have evacuation routes and primary and alternate assembly areas. The assembly area chosen depends on the wind direction and how close the hazard is to the assembly area. Once personnel gather at the assembly area, an accounting (head count) is made. This is to verify everyone has exited the unit or building. If someone is missing, an emergency response crew will be sent into the building to locate them.

CAER and Responsible Care

Community Awareness & Emergency Response (CAER) was launched in 1988 by the Chemical Manufacturers Association (CMA) to respond to public concerns about the manufacture and use of chemicals. The purpose of CAER is to assure site emergency preparedness and to foster community right-to-know. It seeks to assure the community that facilities that manufacture, process, use, distribute, or store hazardous materials initiate and maintain a community outreach program that includes communications training with the public concerning safety, health, and environmental issues. The program fosters a continuing dialogue with a community to respond to questions and concerns about safety, health, and the environment.

Table 12-1 Responsible Care Guiding Principles

Responsible Care Guiding Principles
1. To seek and incorporate public input regarding our products and operations.
2. To provide information on health or environmental risks and to pursue protective measures for employees, the public, and other key stakeholders.
3. To work with our customers, carriers, suppliers, distributors, and contractors to foster the safe use, transport, and disposal of chemicals.
4. To support education and research on the health, safety, and environmental effects of our products and processes.
5. To lead in the development of responsible laws, regulations, and standards that safeguard the community, workplace, and environment.

© Cengage Learning 2013

The **Responsible Care** program is the petroleum and refining industry's effort for minimizing chemical dangers while maximizing public approval of the industry. Though the public may not be as skeptical of the industry today as it was in the past the original goal of building public trust in the industry has not quite been met. Although progress has been made on issues such as toxic releases and workplace injuries, the chemical industry's image still ranks low in public surveys. Responsible Care's guiding principles are shown in Table 12-1.

Cardinal Rules

Cardinal Rules are very important rules, which if broken, have severe disciplinary consequences, such as time off without pay or termination of employment. Examples of Cardinal Rules of a civil society are "Thou shall not kill" and "Thou shall not steal." An infraction of a processing cardinal rule can cause a major environmental or safety incident. Cardinal rules are not required by any regulatory agency, but are considered a best management practice in the process industry. Examples of some common cardinal rules in use throughout most industries are bulleted below. These are only a few of the more common rules. Each manufacturing site will have its own set of cardinal rules.

- No bringing alcoholic beverages onto a workplace or consuming alcoholic beverages while at the workplace.
- No using drugs or being under the influence of drugs within the boundaries of the company.
- No possessing firearms within the boundary of the plant site either in your personal vehicle or on your person.
- No fighting with another employee or threatening violence to another employee.
- No possessing matches, lighters, or smoking in restricted areas.

Mutual Aid Agreements

Mutual aid agreements are agreements between site managers and local governments to render aid in case of need. All manufacturing plants of significant size that deal with hazardous materials have their own fire departments and trained first-aid personnel. They usually have at least one ambulance and one fire truck. Where two or more of these type plants are located fairly close together, each site's emergency response coordinators develop a mutual aid agreement to render aid in the event of a release or other hazardous situation at either site. This is good common sense and basic teamwork. If your site has a large fire and your fire department has only one fire truck, a radio call can bring more trucks and fire fighters from other sites or the nearest municipality to help control and douse the fire.

Hazard Communication Program

Each work site that requires its employees to handle chemicals must write and administer a Hazard Communication Program. The program's intent is to ensure that the hazards of all chemicals produced or imported are evaluated and information concerning their hazards is transmitted to the employee. We will cover this program in more detail in a separate chapter.

Buddy System

Management determines where the buddy system is used. Some tasks are so hazardous lone individuals should never attempt them. The buddy system is employed to monitor and safeguard persons who undertake hazardous operations. There are two common buddy system methods. In the first, two persons constitute a buddy pair and are subjected to the same hazard at the same time and under the same conditions. Each must look out for the well-being of the other, monitoring the other's activities, or providing assistance when required. Power company personnel who work on live high-voltage electrical systems use this type of mutual aid and surveillance.

In the second type of buddy system, only one of the pair is exposed to the hazard. The other's sole duty is to protect and assist the person in danger, should the need arise. A common example in industrial work occurs during confined space entry. When a worker must enter a tank for cleaning or repair purposes, a buddy is stationed outside to monitor the well-being of the person inside. The buddy may provide warnings of any adverse condition noted, assist the worker if aid is needed, or call for outside assistance. The outside buddy in this system should have no duty except that of monitoring the worker in danger.

Warning Means and Devices

Warning means and devices (see Figure 12-2) are means of avoiding accidents by attracting or focusing the attention of the employee on a hazard. Warnings are required by law to inform workers, users, and the public about any dangers that might not be obvious. Every method of identifying and notifying personnel that a hazard exists requires communication. All human senses have been and are used for this purpose. Warning means and devices consist of:

- Visual devices (signs, flashing lights, flags, etc.)
- Audible devices (alarms, horns, klaxons, etc.)
- Odorants (natural gas odorant)

Vision is man's principal sense, and signs and labels are the prime means of transmitting information of the existence of hazards. Throughout processing industries large black-and-yellow signs or bright red signs draw attention and indicate the presence of a hazard. However, signs consisting merely of words are inadequate. More and more, signs are combining symbols with a few key words, and often, the words are in more than one language. Tens of millions of Americans are functionally illiterate and may not be able to read a label on a bottle. A stockyard worker destroyed a herd of cattle because he didn't understand the word "POISON" on a bag he thought was feed. If the symbol of a skull and crossbones had been on the

© Cengage Learning 2013

Figure 12-2 Warning Signs on Analyzer Building

bag next to the word "poison," the worker might have understood he was not handling a feed bag. Billions of dollars in damage occurs annually because workers can't read instructions.

Personal Protective Equipment (PPE)

Personal protective equipment is the equipment persons wear for protection against hazards that cannot be engineered out of the system. The need for personal protective equipment can be divided into three categories:

1. For a scheduled hazardous operation, such as sample collecting (see Figure 12-3). In this case protective gloves, splash shield and other PPE might be required.

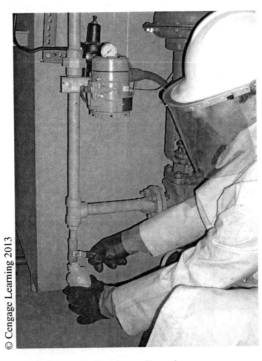

© Cengage Learning 2013

Figure 12-3 Process Employee Collecting a Sample

2. For investigative and corrective purposes it may be necessary to determine whether an environment has become dangerous because of a chemical leak. An air pack and chemical suit might be required in this instance.
3. The severest requirement for PPE may be due to accidents. The first few minutes after an accident occurs may be the most critical and reaction time to suppress or control any injury or damage is extremely important. Because of this, protective equipment must be simple and easy to don and operate.

Though most employees understand the need for PPE, many neglect the use of it due to laziness, because they are willing to take a risk, or because they consider the PPE too uncomfortable to wear. Management must have a proactive PPE program and enforce this method of hazard control. Programs that are not enforced are not programs.

ADMINISTRATIVE ACTIVTIES

Administrative controls can be broken down into the two categories of programs and activities. We have just looked at some programs, which are the written documents that explain how hazards are to be controlled. Administrative activities are the programs put into action. Administrative hazard control activities can consist of any of the following:

1. Training—providing instruction on job tasks, skills, and knowledge.
2. Housekeeping—maintaining a clean, neat, orderly, and safe workplace.
3. Permit System—preparing a job site to ensure conditions are safe for a specific task to begin.
4. Inspections (safety)—conducting regular, scheduled checks to ensure a safe and healthy workplace.
5. Audits (safety)—conducting regular and scheduled checks for safety, health, and environmental compliance.
6. Investigations (safety)—taking action to determine the cause or causes of an incident or accident.
7. HAZOP—conducting a hazard and operability study to identify any possible flaws in a system, operation, or piece of equipment that could lead to an accident.
8. Industrial Hygiene Monitoring—taking samples in the worker's breathing zone to determine if toxic or hazardous substances are present at unacceptable levels.
9. Fugitive Emissions Monitoring—testing valves and flanges to determine if leaks are allowing toxic or hazardous substances to be present at unacceptable levels that are dangerous to workers and polluting to the environment.
10. Safe Work Observations—looking at selected work practices on a regular basis and providing feedback on the safe and potentially unsafe acts observed.

Administrative activities require management support and commitment of budget for expenses that include:

- Overtime costs for people in training
- Expended material costs, such as refilling fire extinguishers after fire training, respirator cartridge replacement, etc.
- Consultant or special schooling fees, such as sending fire brigade members to firefighting schools

Training and Information

Training and information dissemination is an important administrative function. Process employees engage in a variety of training activities throughout their careers. Their training can be broken down into three major categories:

- **Formal education**—many process industry employers now expect new employees to possess more than a high school degree. In some cases, an employee can no longer retain their current job unless they have a college degree. They may be required to take college courses and work toward an associate's degrees or four-year degree because the better educated their workforce, the more skilled and flexible it becomes.
- **Plant and unit-specific training**—as well as on-going safety, health, and environmental training. Process employees are tested and often certified, on the knowledge and skill acquired as a result of training. This training is periodic and determined by company and federal regulations.
- **Information**—to keep all employees knowledgeable of changes occurring in the workplace, corporate guidelines or benefits, employers constantly provide employees with information via memos, updated procedures, email messages, short "town hall" meetings, etc. Employees are responsible for acquiring and retaining this information.

Housekeeping

Housekeeping in the process industry is basically the same as housekeeping in the home. In order to maintain a safe, orderly environment workers must keep their work areas (control room and outside unit area) neat, clean, and uncluttered. Good housekeeping accomplishes the following goals:

- Prevents accidents by promoting a safe and orderly work environment
- Saves money by preventing accidents and increasing morale
- Increases worker productivity by workers not having to remove equipment and material in their way or go out of their way to avoid congested areas
- Improves worker morale just by having a neat, orderly workplace

Safe Work Permits

Safe work permits, also called *permits,* give permission to plant personnel to perform work and maintenance after they have satisfied certain requirements listed on the permit. The function of a permit or permit system is to ensure that appropriate precautions are taken to prevent harm to personnel about to engage in a potentially dangerous activity, such as enter an enclosed space or weld in the middle of the process unit. It also releases custody of the equipment to the workers about to make the repair. Permits require special forms, and in some cases, signs, tags, and other associated equipment or material. Each plant has its own permit system. A later chapter will take a look at three very common permit systems—lockout/tagout, hot work, and confined space. The way a permit helps prevent accidents and incidents is by listing the steps required to complete a task, assessing the hazards associated with each step, eliminating or minimizing the threats, and maintaining safe working conditions.

Inspections

Inspections involve conducting regular safety, health, and/or environmental checks. Inspections are proactive because they are conducted prior to a need for action. Inspections keep workers on their toes because they do not have to be announced ahead of time, though they often are. Good managers and supervisors make a habit of walking through their area of responsibility frequently and inspecting for things that might cause an incident or accident. Inspections are usually conducted locally (by plant personnel) to insure that site safety, health, and environmental procedures are being followed. They are usually conducted on a regular schedule, such as weekly, monthly, quarterly, or yearly. Inspections involve the use of checklists which contain a list of inspection items. Often these checklists incorporate rating scales that the inspector uses to rank the unit's or plant's performance in complying with an inspection item. Process employees will often be required to assist or conduct inspections within their unit or plant. One of the more common inspections is of emergency response equipment (fire extinguishers, fire hoses, fire blankets, safety showers, etc.).

Audits

Audits, like inspections, involve conducting regular safety, health, and/or environmental checks. They are typically conducted by outside personnel—by a corporate entity, a sister company, or a consultant—to determine if the plant is in compliance with company or regulatory requirements. Corporate audits are also done according to a schedule, but tend to be done less frequently than routine safety inspections. Audits also involve the use of checklists, which contain a list of items to be audited. The primary reason for a company to conduct a health and safety audit is to ensure regulatory compliance. Violations of Occupational Safety and Health Administration (OSHA) requirements can result in significant penalties and poor publicity for the organization. Other reasons for conducting a health and safety audit include risk management, financial liability, proactive operations, and increasing health and safety awareness.

Interpersonal skills are very important during audits because you are finding fault with someone else's work. It is necessary to be honest and direct when explaining areas of noncompliance. In most cases, noncompliance issues are a result of an employee not doing their job correctly because they weren't aware of the requirements. Audits rarely result in disciplinary action but do result in more training and more work to correct the noncompliances because simply identifying areas of noncompliance does not solve the actual reason for the occurrence of the noncompliance. The audit provides the basis for the unit or work site to conduct a root cause analysis in order to eliminate the noncompliance.

An example of noncompliance in an audit of a petrochemical facility is a respirator found covered with dust and lying on a tabletop in a sample room. This is in violation of the Respiratory Protection Standard (29 CFR 1910.134), which requires that respirators be stored in a convenient, clean, and sanitary location. This was a case of improper respirator storage. A root cause analysis would determine if operators know where and how respirators should be stored. If they don't, then additional respirator training may be required.

Accident Investigations

Many accidents appear to have simple, easily determinable causes; but changes in legal aspects may require competent investigative abilities. For example, an accident occurs in which a chain breaks under a load. The load drops on a worker and kills him. At one time, the report of investigation may have described it in fairly simple terms and indicated that a contributory cause of the worker's death was an unsafe act on his part, which was to put himself in a dangerous position. Reports of a later period might also include as a contributory cause supervisory error because the supervisor should not have permitted the worker to put himself in danger. However, today we live in a litigious world. An attorney for the worker's dependents might seek to prove the supervisor knew the worker was in the dangerous position or knew the chain was bad and permitted the worker to continue with the operation and so was guilty of gross negligence. The attorney might also bring suit for the dependents against the manufacturer of the chain, claiming negligence in its manufacture. For this, analysis of the chain and the part where it broke might be required. The chain manufacturer might claim as a defense that the chain had been damaged through long use or had been used improperly in that it was too small for the intended load. Investigations by representatives of different organizations with different interests will attempt to ensure that whomever it represents is not held to be at fault. The accident will be very expensive for one or more parties.

Investigations are considered to be reactive because they are conducted in response to an accident. Investigations involve taking action to determine the cause or causes of an incident or accident. Usually, investigations do not take place unless one of three things has occurred:

1. An incident—a hazardous situation where no injury or process upset occurred.
2. An accident—a hazardous situation where the result was an injury, equipment damage or process upset.
3. A near miss—an unsafe act without incident or accident.

Investigations are generally performed by a team of individuals. The investigative team members may not all be from the same unit or plant division. This is deliberate so that people with different viewpoints and areas of expertise can look at the same problem and bring different insights to the problem. Investigations are conducted similarly to the way you would go about problem solving. The investigative team would take the following steps:

1. Decide the problem to be addressed.
2. Arrive at a statement that describes the problem: what happened, when it happened, the extent of the problem, etc.
3. Develop a complete picture of all the possible causes or the problem and determine the root cause of the problem.
4. Develop an effective solution and action plan.
5. Implement the solution.
6. Confirm expected results.

Process employees are often required to participate in investigations. They may be called upon to participate because they were directly involved in the accident or are familiar with the equipment or procedures that apply to the accident. Their familiarity and expertise make them valuable team members in an investigation.

HAZOP (Hazards and Operability Study)

There are many different methodologies under the Process Hazards Analysis (PHA) umbrella. Depending on the goals of the study and the time available, one of several different studies would be completed. One of the most comprehensive PHAs is a **HAZOP (Hazards and Operability Study)**, which identifies operability problems as well as hazards to personnel, company property, and the environment.

To prevent the undesirable consequences of accidents, hazards that can lead to accidents must be identified. Effective hazard identification and control require a systematic, comprehensive, and precise analysis of the process system and its operation. Once hazards are recognized, the adequacy of the mechanisms to control the hazards is evaluated. If control mechanisms do not exist or are inadequate, this must recognized and actions recommended. Several of the benefits of using a HAZOP are the:

- Identification of hazards to personnel, facilities, and the environment
- Prioritizing risk reduction
- Mitigation of legal liability
- Reduction of operating costs
- Improvement of employee morale
- Shortening of project schedules
- Optimization of productivity and profitability

In regard to the last bulleted item, it has been estimated that approximately 70 percent of the recommendations from a HAZOP study relate to improving operability. If you improve operability, you improve profitability.

Industrial Hygiene Monitoring

Industrial hygiene monitoring involves sampling of the working environment to determine if any hazardous agents are present at unacceptable levels (levels determined by OSHA). Industrial hygiene monitoring is conducted by industrial hygienists using a variety of tools to capture data on workplace hazards and analyze this data and compare it to published standards. Common industrial hygiene sampling activities include (1) noise monitoring, (2) toxic substances sampling, and (3) ergonomic studies. Many hazardous fluids at very low levels may not be detectable by smell or be covered up by a background smell that is not hazardous. A process employee may unwittingly be exposed to a hazard. Besides being harmful to process employees these emissions harm the environment and may result in monetary penalties when discovered.

Detecting flammable or toxic emissions can be accomplished either by fixed sensors or by personnel using portable gas detectors. Most, if not all process technicians, are trained to use them. Another way is by process employees wearing detectors or badges. If the detector senses a hazardous compound at unacceptable levels, it will alert the employee wearing it with some type of alarm(s). Badges may change color when exposed to an unacceptable level of a compound. Other types of badges consist of activated charcoal or some absorbent that will absorb the volatile compounds or gases in the air. The badge is worn for a specific time period, then collected and sent to a lab that will extract the absorbed compounds and analyze them to determine the concentration in that area.

Fugitive Emissions Monitoring

Fugitive emissions monitoring is similar to industrial hygiene monitoring but more rigorous and often contracted out due (see Figure 12-4) to its extensive requirements. Emissions monitoring involve testing of piping, valves, and other equipment with portable gas detectors or other means to determine if volatile organic compounds (VOCs) and/or other specific compounds are leaking into the environment. Emissions are reported to regulatory agencies. Regulatory agencies are very concerned about fugitive emissions because one plant with extensive piping and thousands of valves may have hundreds of minor leaks releasing chemicals into the local atmosphere. If there are 1000 plants with hundreds of leaks each, then there is a serious potential that the local air quality will be affected, causing human health problems, smog, elevated ozone levels, etc.

Safe Work Observations

Safe work observation (SWO) is a behavior-based observation and feedback process. It is conducted with the use of a standardized form developed by each manufacturing site or they may be able to purchase a generic form. The person conducting the SWO walks through a work area and takes note of what they see occurring. It isn't uncommon to perform an SWO and have no negative findings. In fact, the use of the SWO usually results in more workers being praised for their safe work habits than constructively criticized for unsafe behavior.

© Cengage Learning 2013

Figure 12-4 Fugitive Emissions Monitoring Tags

As a rule, no discipline results from filling out a SWO. The entire intent of this safety function is keep people aware of safety. A typical example of a finding from a safe work observation might be a worker was noticed collecting a sample without wearing chemical resistant gloves. After the SWO is completed, a good comment to make to the worker who was not wearing proper PPE would be, "Joe, I saw you collecting a sample of ethyl benzene without wearing chemical resistant gloves. Exposure to ethyl benzene can cause cancer. You've got three children at home. If you want to live to see your grandchildren, please wear the gloves you are supposed to wear." This is a polite but impressive way to remind workers of the importance of safety.

SUMMARY

After engineering controls are implemented administrative controls are the next step in hazard reduction. Administrative controls are procedures put into place to limit employee exposure to hazards. Their function is to influence worker behavior to do what they are supposed to do. Enforcement of the procedures may have penalties that vary in severity but can include termination of employment.

Administrative controls can be broken down into two broad categories: programs and activities. Programs are the written documents that explain how hazards are to be controlled. Activities are the programs put into action. Some of the more common administrative hazard control programs include policies, procedures, and plans. Some of the more common administrative hazard control activities are training, housekeeping, the permit system, and safety inspections and audits.

REVIEW QUESTIONS

1. _____ are procedures put into place to limit employee exposure to hazards.

2. (T/F) When hazards cannot be engineered out of a workplace then administrative controls are the next logical step for hazard control.

3. The two categories of administrative controls are _____ and _____.

4. List six examples of administrative controls.

5. _____ are programs put into action.

6. (Choose two) Employees are trained on administrative controls:
   a. When first hired
   b. Annually
   c. On a determined periodic basis
   d. When accidents become to frequent

7. The purpose of _____ is to assure site emergency preparedness and to foster community right-to-know.
   a. Responsible Care
   b. Community Awareness and Emergency Response

c. Mutual Aid

d. Evacuation and Accountability plans

8. The _____ program is the petroleum and refining industry's effort for minimizing chemical dangers while maximizing public approval of industry.

a. Responsible Care

b. Community Awareness and Emergency Response

c. Mutual Aid

d. Evacuation and Accountability plans

9. (T/F) Cardinal Rules are not required by any regulatory agency but are considered best management practices.

10. Describe two cardinal rules.

11. _____ are agreements between site managers and local governments to render aid in case of need.

12. _____ are required by law to inform workers, users, and the public about any dangers that might not be obvious.

13. List five administrative hazard control activities.

14. _____ are programs put into action.

a. Training programs

b. Mutual Aid agreements

c. Administrative activities

d. Evacuation plans

15. In order to maintain a safe, orderly work environment workers must keep their control room and outside unit area neat and clean. This activity is called _____.

16. (T/F) Inspections are proactive because they are conducted prior to a need for action.

17. Explain the purpose of safe work permits.

18. An _____ is a hazardous situation where no injury or process upset occurred.

a. Accident

b. Incident

c. Near miss

19. An _____ is a hazardous situation where the result was an injury, equipment damage, or process upset.

a. Accident

b. Incident

c. Near miss

EXERCISES

1. Make a list of ten or more administrative controls used in everyday life (parents with children, law enforcement, transportation, schools, etc.) and explain what hazard they safeguard against.

2. Accountability to the community (neighborhoods) around a processing site is a serious responsibility. Write a one page report on either of the following: **Community Awareness & Emergency Response (CAER)** or **Responsible Care.**

RESOURCES

www.wikipedia.org

www.osha.gov

www.YouTube.com

www.toolboxtopics.com

CHAPTER 13

Personal Protective Equipment (PPE)

Learning Objectives

Upon completion of this chapter the student should be able to:

- *Explain the purpose of personal protective equipment (PPE).*

- *List five types of PPE.*

- *Explain how the level of PPE protection is chosen.*

- *Describe the PPE required for levels A, B, C, and D protection.*

- *Describe at least one condition that would require levels A and B protection.*

- *Describe four different types of gloves used for hand protection.*

- *Explain the difference between goggles and safety glasses.*

- *Explain what is meant by the permeation of protective material.*

- *Explain what is meant by the degradation of protective material.*

- *Describe the protection offered by hard hats.*

- *Describe the protection offered by fire retardant clothing.*

- *Describe three checks to be made on ladders before using them.*

- *Describe the equipment used in fall protection.*

INTRODUCTION

Personal protective equipment (PPE) is last in the hierarchy of hazard control. PPE is less desirable than engineering and administrative controls for the control of hazards but is still critical. The appropriate PPE fitted correctly is a reliable barrier against known hazards. The biggest drawback against PPE is that some workers are careless about their selection of PPE for a job and choose ineffective PPE or they do not ensure that it is properly fitted and used. The function of PPE is to protect the user's entire body, including the respiratory system, eyes, hearing, head, hands, etc.

Examples of PPE commonly used are the:

- Hard hat
- Full-face splash shield
- Eye protection (goggles, safety glasses, full-face shield, etc.)
- Hearing protection (ear plugs, ear muffs, etc.)
- Gloves (chemical resistant, heat resistant, abrasion, etc.)
- Coveralls, apron, splash suit, or fully encapsulated suit
- Boots (often steel-toed)
- Respirators
- Fire resistant clothing (FRC)
- Radios
- Harness and lanyard

The obvious function of these garments and devices makes their use fairly straightforward. However, a simplistic and uneducated approach to using personal protection can result in unsuspected exposures. This is because there are significant complexities in both design and function of the protective devices used to reduce exposure. The worker must be knowledgeable about the limitations of each piece of PPE. Respirators and gloves, of which there are multiple types, used improperly or chosen in ignorance can lead to exposures nearly as high as that of being completely unprotected.

THE PERSONAL PROTECTIVE EQUIPMENT PROGRAM

Management is aware a PPE program may reduce the efficiency and productivity of workers because they are uncomfortable wearing PPE. A good personal protection program requires that management walk through the work areas, and where workers are in non-compliance, enforce their program. Other important aspects of hazard control are adequate housekeeping and proper disposal of contaminated PPE.

There are many suppliers and manufacturers of PPE. Several factors should be evaluated before choosing a supplier. Those factors are the product's quality, whether the product meets the Occupational Safety and Health Administration (OSHA) or the National Institute for Safety and Health's (NIOSH) standards, and availability and delivery of the PPE. Prior to choosing a supplier, test performance data should be requested and samples obtained for circulation among the workforce to get feedback from the actual users. The worst thing management can do is buy bulky or uncomfortable PPE that the workers will avoid wearing, especially when they know comfortable PPE is available.

OSHA and NIOSH always emphasize that the best approach to minimizing a risk or hazard is through engineering design or mitigate it through administrative controls. In real life, however, we can't design out all hazards. Since we can't eliminate the hazard, we must plan to deal with it safely. The way to do that is to create a protective barrier between the worker and the hazard. PPE is that protective barrier. Workers must wear protective equipment when activities will involve (or are suspected of involving) known or suspected atmospheric contamination; when vapors, gases, or particulates may be generated by work activities; or when direct skin contact with toxic or irritating chemicals may occur.

LEVELS OF PPE PROTECTION

Equipment to protect the body against contact with known or anticipated toxic chemicals has been divided into four categories according to the degree of protection afforded:

- **Level A** should be worn when the highest level of respiratory, skin, and eye protection is needed.
- **Level B** should be worn when the highest level of respiratory protection is needed, but a lesser level of skin protection.
- **Level C** should be worn when the criteria for using air-purifying respirators are met.
- **Level D** should be worn only as a work uniform and not on any site with respiratory or skin hazards. It provides no protection against chemical hazards.

The **level of protection** selected should be based on (1) the type and measured concentration of the chemical substance in the ambient atmosphere and its toxicity, and (2) the potentials for exposure to substances in air, splashes of liquids, or other direct contact with material while doing work. In situations where the type of chemical, concentration, and possibilities of contact are not known, the appropriate level of protection must be selected based on professional experience and judgment until the hazards can be better identified. This may require a person to suit up in a Level A or B PPE and investigate/test the atmosphere/area to determine the hazard level.

Level A Protection

This is the highest level of breathing and skin protection. In other words, it is whole body protection. Typical personal protective equipment consists of:

- A supplied-air respirator approved by the Mine Safety and Health Administration (MSHA) and/or NIOSH. Respirators may be pressure-demand, self-contained breathing apparatus (SCBA) or a pressure-demand, airline with escape bottle for immediately dangerous to life and health (IDLH) or potential IDLH atmosphere.
- A fully encapsulating chemical-resistant suit. This suit literally is an envelope surrounding the entire body with inserts for feet, hands, and fingers.
- Gloves (inner), chemical-resistant.
- Boots, chemical-resistant, steel toe, and shank.
- A hard hat (optional and worn under the fully encapsulating suit).
- A radio (inherently safe).

When would Level A protection be worn? What follows is a partial list of conditions that warrant suiting up in PPE to Level A protection.

- The chemical substance is known and requires the highest level of protection for skin, eyes, and respiratory system based on measured (or potential) high concentrations of atmospheric gases or particulates.
- The task involves a high potential for splash, immersion, or exposure to unexpected vapors, gases, or particulate material highly toxic to the skin.
- Substances with a high degree of hazard to the skin are known or suspected to be present and skin contact is possible.
- Visible fluids are leaking from containers.
- Direct readings on portable gas detectors indicate high levels of various unidentified vapors and gases in the air.

Fully encapsulating suits are primarily designed to provide a gas or vapor tight barrier between the wearer and atmospheric contaminants. Therefore, Level A (see Figure 13-1) is generally worn when high concentrations of airborne substances are known or thought to be present and these substances could severely affect the skin. Since Level A requires the use of a self-contained breathing apparatus, the eyes and respiratory system are also more protected. Suiting up in Level A PPE requires two people: the employee who will wear the suit and a helper to assist them in donning the equipment. The helper must keep an eye on the wearer as they put on the SCBA and ensure the SCBA air supply is on before being sealed in the encapsulating suit.

© Cengage Learning 2013

Figure 13-1 Employee Wearing Level A Protection

Level B Protection

Personal protective equipment required for Level B protection includes:

- Supplied-air respirator (MSHA/NIOSH approved). Respirators may be pressure-demand, self-contained breathing apparatus or pressure-demand, airline respirator with escape bottle for IDLH or potentially IDLH atmospheres.
- Chemical-resistant clothing (one or two-piece chemical-splash suit or disposable chemical-resistant one-piece suits)
- Gloves (outer), chemical-resistant
- Gloves (inner), chemical-resistant
- Boots (outer), chemical-resistant, steel toe, and shank
- Hard hat
- Radio (inherently safe)

Some of the conditions that would warrant suiting up for Level B protection are:

- The type and atmospheric concentration of toxic substances has been identified and requires a high level of respiratory protection, but less skin protection than Level A. Typically, these would be atmospheres with IDLH concentrations, but the substance or its concentration does not represent a severe skin hazard or the atmosphere does not meet the selection criteria permitting the use of air-purifying respirators.
- The atmosphere contains less than 19.5 percent oxygen.
- It is highly unlikely that the work being done will generate high concentrations of vapors, gases, or particulates, or splashes of materials that will affect the skin of personnel wearing Level B protection.
- Atmospheric concentrations of unidentified vapors or gases are indicated by direct readings on instruments but the vapors and gases are not suspected of containing high levels of chemicals harmful to the skin.

Level B protection (see Figure 13-2) does not afford the maximum skin and eye protection as does Level A. Chemical resistant clothing is not gas, vapor, or particulate tight, whereas a fully encapsulated suit is. Level B does provide a high level of protection to the respiratory tract. The productivity and comfort point of Level B is vastly superior over Level A.

Level C Protection

Personal protective equipment for a Level C environment consists of:

- Air-purifying respirator, full-face, canister-equipped, and MSHA/NIOSH approved
- Chemical-resistant clothing, such as chemical splash suit, chemical resistant hood and apron, and disposable chemical-resistant coveralls
- Gloves (outer), chemical-resistant
- Boots (outer), chemical-resistant
- Hard hat
- Radio (inherently safe)

Figure 13-2   Employee Wearing Level B Protection

Level C protection is distinguished from Level B by the equipment used to protect the respiratory system, assuming the same type of chemical-resistant clothing is used. The main selection criterion for Level C is that conditions permit wearing air-purifying respirators. An environment that would warrant donning Level 3 PPE would have the following characteristics:

- Oxygen concentration is not less than 19.5 percent by volume.
- Measured air concentrations of identified substances will be reduced by the respirator below the substance's threshold limit value (TLV) and the concentration is within the service limit of the canister.
- Atmospheric contaminant concentrations do not exceed IDLH levels.
- Atmospheric contaminants, liquid splashes, or other direct contact will not adversely affect any body area left unprotected by chemical-resistant clothing.
- Job functions do not require a self-contained breathing apparatus.

Level D Protection

Level D protection requires minimum PPE. Level D protection is essentially a work uniform that can be worn only in areas where there is no possibility of contact with contamination. Many people say Level D is no protection at all. This is true. The normal work uniform—fire resistant clothing (FRC), safety glasses with shields, and safety shoes—is

Table 13-1 Levels of Protection

Level	Respirator	Chemically Protective Clothing	Protection
A	Pressure-demand full facepiece SCBA or pressure-demand supplied-air respirator with escape SCBA	Fully-encapsulating chemical resistant suit	The highest level of respiratory, skin and eye protection
B	Pressure-demand full facepiece SCBA or pressure-demand supplied-air respirator with escape SCBA	Chemically-resistant clothing, such as one-piece disposable suit or overalls and long-sleeved jackets	The same level of respiratory protection but less skin protection
C	Full facepiece air purifying canister equipped respirator	Chemically-resistant clothing, such as one-piece disposable suit or overalls and long-sleeved jackets	The same level as skin protection as level B but a lower level of respiratory protection
D	No respiratory protection required	FRC, safety shoes/boots, safety glasses, hard hat	No respiratory protection, minimal skin protection

© Cengage Learning 2013

considered Level D protection. Most work in normal plant operations can be safely performed using Level D protection. The PPE required for this level is:

- Coveralls or fire-resistant clothing (FRC)
- Boots/shoes, leather or chemical-resistant, steel toe, and shank
- Safety glasses
- Hard hat
- Radio (intrinsically safe)

Meeting any of the following criteria allows use of Level D protection:

- No airborne contaminants are present
- Work functions preclude splashes, immersion, or potential for unexpected inhalation of any chemicals

We have looked at the four levels of PPE (see Table 13-1) and might have found them to be slightly confusing. Let us say you and a coworker can't decide what level to don in a particular situation. There is an easy answer to your concern: **when in doubt, always use the higher level of protection.**

CHEMICALLY RESISTANT SUITS

There is no universal chemically resistant suit that is safe to use with all chemicals. Obviously, the suit material will depend on the type of chemical exposure involved. Since it fully encapsulates, it also involves gloves (both sides, inner and outer) which are chemical resistant and chemical resistant boots. When selecting suit material, it is best to seek the aid of the vendor. They know what they make and what it can and can't withstand. Suit material can be subject to (1) **permeation**, (2) **degradation**, and (3) **penetration** by a specific chemical(s).

- Permeation is the amount of chemical that will pass through a material in a given area in a given time. Though zero permeation is desired, it may be impossible or economically prohibitive. Different manufactures use different rate units of permeation. If permeation is very slow a worker can enter the environment, complete their task, and leave with minimum or no exposure.
- Degradation is the gradual chemical destruction of a material. A suit may be perfectly sealed but if the material doesn't have adequate resistance to the chemical, the chemical can react with the suit material and begin to destroy it. Then it is only a matter of time before the chemical contacts flesh.
- Penetration applies to the risk that the chemical can go through the garment material, or through seams and other imperfections.

A familiar complaint of workers who must don Level A and B PPE is that the equipment is very uncomfortable (hot and sweaty) and cumbersome, which lowers their productivity. The use of protective clothing and respirators increases physical stress, in particular, heat stress. Wearing protective equipment also increases the risk of accidents. It is heavy, cumbersome, decreases dexterity and agility, interferes with vision, and is fatiguing to wear. New suits are being developed that are comfortable and impose less stress on workers. PPE selection is a process that involves a mix of experience, common sense, and quantitative risk assessment.

Acid Suits, Aprons, and Smocks

Acid suits (see Figure 13-3) are designed to resist corrosive materials, such as hydrochloric and sulfuric acids and caustics, like sodium hydroxide. Process operators have many

© Cengage Learning 2013

Figure 13-3 Acid Suit and Splash Shield

responsibilities that involve them with acids or caustics, such as loading or unloading railcars of acids and caustics. Acid suits are like heavy rain gear, and on a warm day make the wearer very hot but the suit forms a barrier between the wearer and splashes of very strong acids and caustics which could result in serious chemical burns. Splash shields are worn in conjunction with the acid suit to protect the technician's eyes. Aprons and smocks protect a major portion of the wearer against splashes.

GLOVES AND HAND PROTECTION

Providing protection that guards the hands is necessary under the OSHA PPE Standard 1910.138. This regulation requires employers to provide protection for employees who may be exposed to hazards that include skin absorption or harmful substances and chemical burns. OSHA lists four broad categories of protective gloves:

- Leather, canvas, or metal mesh that provides protection against cuts, abrasions, and burns. Leather and canvas also provide protection against sustained heat.
- Fabric or coated fabric gloves made of cotton or other fabrics that protect against dirt, chafing, abrasions, and splinters.
- Chemical and liquid resistant (discussed later in detail)
- Insulating rubber gloves that protect against extreme cold.

Hand protection is available to protect against cut/punctures, abrasions, thermal burns, vibration, chemical exposures, and electrical shock. Gloves used for hand protection (see Figure 13-4) also protect the user's health when handling chemicals because many chemicals are acutely or chronically toxic. There is not a single glove that will protect from all hazards. Selection of gloves must be based on the hazards that are present, the job task, work conditions, and the duration of use. Gloves to be used to protect against the effects of chemical use should be selected based on each manufacturer's glove selection charts. Do not assume that the protection offered by one manufacturer's

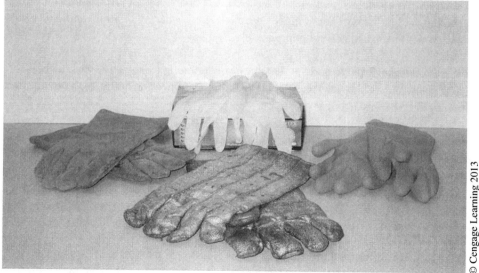

© Cengage Learning 2013

Figure 13-4   Types of Gloves

glove will apply to all types of similar gloves. The protection of each glove is based on the manufacturing processes and glove thickness. You may be assigned to a process unit with a variety of chemicals and temperatures. Because of that variety the unit will maintain a supply of five to eight different types of gloves. As a new hire, you will have to learn which glove to use for which task, otherwise you will harm yourself. *Remember, all gloves are not created equal.* Chemical concentration and the temperature of the chemical affect the permeation rate of a chemical through glove material. This is why it is very important to utilize the glove selection guides provided by the manufacturer. Refer to the ORCBS Glove Guide for assistance in glove selection, *www.orcbs. msu.edu/chemical.*

Glove Use and Maintenance

Gloves that are torn or damaged should not be used. Consideration of the following items is necessary when using gloves to protect against chemical hazards: Determine that the glove will provide adequate protection for the chemical to be encountered. If multiple chemical hazards exist, base the effectiveness of the glove on the chemical with the fastest breakthrough time. By *breakthrough* it is meant how long it takes for the chemical to permeate through the glove material and make contact with the skin. Most gloves are resistant to permeation for a certain amount of time. Inspect gloves prior to each use. The only gloves allowed to be kept and used the next day(s) were work gloves for handling rough, hot, or cold materials, but never chemicals.

Chemically Resistant Gloves

Chemically resistant gloves are made from rubber (latex, nitrile, or butyl) or a synthetic composition such as neoprene. Frequently used gloves are:

Butyl Rubber Gloves—Provide protection from nitric acid, sulfuric acid, hydrofluoric acid, red fuming nitric acid, rocket fuels, and peroxide. These gloves have a high impermeability to gases, chemicals, and water vapor, and resistance to oxidation and ozone attack. They have high abrasion resistance and remain flexible at low temperatures.

Natural Latex or Rubber Gloves—Provide protection from most water solutions of acids, alkalis, salts, and ketones, plus, they are resistant to abrasions occurring in sandblasting, grinding, and polishing. These gloves have excellent wearing qualities, pliability, and comfort and are a good general-purpose glove.

Neoprene Gloves—Provide good protection from hydraulic fluids, gasoline, alcohols, organic acids, and alkalis. They have good pliability and finger dexterity, high density, and tensile strength, plus high tear resistance.

Nitrile Rubber Gloves—Provide protection from chlorinated solvents (trichloroethylene, perchloroethylene, etc.). They are intended for jobs requiring dexterity and sensitivity, yet they stand up under mechanical use even after prolonged exposure to substances that cause other glove materials to deteriorate. They also resist abrasion, puncturing, snagging, and tearing.

HEAD PROTECTION

OSHA's standard for head protection, 29 CFR: 1910.135, requires that the employer ensure that each affected employee wears a protective helmet when working in areas where there is a potential for injury to the head from falling objects. The employer shall ensure that a protective helmet designed to reduce electrical shock hazard is worn by each such affected employee when near exposed electrical conductors which could contact the head. Protective helmets (hats) protect in the following ways:

- A rigid shell that resists and deflects blows to the head
- A suspension system inside the hat that acts as a shock absorber
- Some hats serve as an insulator against electrical shocks
- Shields the scalp, face, neck, and shoulders against splashes, spills, and drips

Hard hats should be stored out of direct sunlight and periodic inspections should be made of all hard hats. Those exposed to sunlight for prolonged periods are susceptible to ultra-violet degradation. At the first sign of a loss of surface gloss or flaking, the hard hat should be replaced. Replace the liner about once a year.

Protective helmets purchased after July 5, 1994 shall comply with ANSI Z89.1-1986, "American National Standard for Personnel Protection-Protective Headwear for Industrial Workers-Requirements," which is incorporated by reference as specified in Sec. 1910.6, or shall be demonstrated to be equally effective. The head is a very delicate and important part of the body that contains the valuable organs for sight, smell, hearing, eating, and speaking (the mouth), and the brain. Injuries to the head are very serious so employees should use their head and wear their hard hat.

Regarding head injuries, a Bureau of Labor Statistics (2009) survey noted that more than one-half of the workers with blows to the head were struck on the head while they were looking down and almost three-tenths were looking straight ahead. While a third of unprotected workers were injured when bumping into stationary objects, such actions injured only one-eighth of hard hat wearers. Where these conditions exist, head protection must be worn to eliminate injury.

FOOT PROTECTION (1910.136)

The use of foot protection minimizes the potential for occupational foot injuries. Terms such as "safety shoes" and "protective footwear" refer to shoes which are specifically designed and tested to withstand stimuli defined in ASTM-F 2413 March 2005. Managers and employees must review their tasks to determine the potential for exposure to foot hazards and the severity of exposure and assure that foot protection is worn when necessary.

Foot hazards may be posed by:

- Falling objects
- Rolling objects
- Sharp material at foot level
- Heavy debris/objects which may be kicked
- Puncture hazards

Safety shoes must be worn if an employee is frequently (at least once/week) exposed to potential foot hazards. This includes employees who must frequently visit potential foot hazard areas such as warehouses, mechanical equipment rooms, construction areas, etc. The shoes must be worn during all foot hazard exposures. Safety shoes must be worn if a task is done infrequently (less than once/week) but the severity of potential injury could result in a broken bone. Typical foot protection for process technicians is leather shoes or boots with steel toes, and in many cases, steel shanks.

FIRE RETARDENT CLOTHING

OSHA requires flame retardant clothing (FRC) for workers at sites based on the quantity of flammable materials (liquids, solids, and gasses) and reactive chemicals that are handled and/or processed with activities at a facility. Both OSHA and NFPA 2113 see the need for FRC if 1) flash fire hazards exist on a continuous basis in various site areas and 2) when employees such as operators and maintenance personnel are in the areas where flash fire hazards exist. FRC can significantly reduce a burn injury by giving the wearer precious escape time from an ignition source and can greatly increase the chance for survival if the wearer is caught in a flash fire or electric arc.

The outermost layer of clothing must always be flame resistant. A worker should not wear flammable clothing over their FRC. Wearing highly-flammable garments, such as nylon parkas, over FR clothing greatly compromises the overall protection of the FRC. Even though the underlying FRC garments will not ignite, the flammable jacket can become a combustible fuel source that can still severely burn the wearer through contact with the flames or heat transfer through the FRC fabric. Ratings of FRC apparel are typically found on the garment label.

Typically, sites lease their FRC from a vendor and workers have five to seven days of FRC available for work. At the end of the workday, they usually place their dirty FRC in a collection site. The vendor collects it, washes it, and returns it to the user work area. This is done to prevent the worker from taking the FRC home and washing it with the wrong detergents, bleaches, etc., that might degrade the fire resistant compounds of the garment.

EYE PROTECTION

Each day, about 2000 U.S. workers have a job-related eye injury that requires medical treatment. About one third of the injuries are treated in hospital emergency departments and more than 100 of these injuries result in one or more days of lost work. The majority of these injuries result from small particles or objects striking or abrading the eye (Center for Disease Control, Prevention and Eye Safety, 2010). Large objects may also strike the eye/face, or a worker may run into an object causing blunt force trauma to the eyeball or eye socket. Chemical burns to one or both eyes from splashes of industrial chemicals or cleaning products are common. Thermal burns to the eye occur as well. Among welders, their assistants, and nearby workers, UV radiation burns (welder's flash) routinely damage workers' eyes and surrounding tissue. In addition to common eye injuries, health care workers, laboratory staff, janitorial workers, animal handlers, and other workers may be at risk of acquiring infectious diseases via ocular exposure. Infectious diseases can be transmitted through the mucous membranes of the eye as a result of direct exposure to splashes bearing infectious organisms.

Types of Eye Protection

The eye protection (see Figure 13-5) chosen for specific work situations depends upon the nature and extent of the hazard, the circumstances of exposure, other protective equipment

© Cengage Learning 2013

Figure 13-5 Types of Eye Protection

used, and personal vision needs. Eye protection should be fit to an individual or adjustable to provide appropriate coverage. It should be comfortable and allow for sufficient peripheral vision. Selection of protective eyewear appropriate for a given task should be made based on a hazard assessment of each activity. Eye protection is available for protecting the eyes from impact, dust, splashes, radiation, laser light, and UV light.

Safety spectacles (glasses)—provide particle protection only. They come with two ratings, high impact and basic impact. If you are working in an environment that puts you at risk of flying objects or any kind of impact risk, your safety glasses must have side shields. Safety glasses side shields can be detachable or permanently attached, in which case they are also known as safety spectacles. In high impact testing, a high velocity test is performed by shooting a quarter-inch diameter steel ball at the lens at a speed of 150 feet per second. To pass, the lens must not crack, chip, or break, and it must not become dislodged from the lens holder.

Safety goggles—form a seal around the eyes to provide complete protection from particles and dust. Goggles are generally the most recommended eye protection on any work site, from construction site to laboratory.

Face shields—are the ultimate choice for eye and face protection. They protect the wearer's entire face from flying particulates and splashes. Face shields are secondary protectors only and must be worn with safety glasses or goggles, as stated in ANSI Z87.1.

HEIGHT AND FALL PROTECTION

Process employees may be required to climb ladders, fixed or portable, and work at heights above six feet. This requires them to have training on height and fall protection. Generally speaking, height protection is working at any height below six feet above the floor or grade. Fall protection is required is working at six feet or higher from the floor or grade.

HEIGHT PROTECTION

To be safe while using a portable ladder, such as a step ladder, employees should follow the guidelines below:

- Ensure the ladder has retained its strength, has no cracks or loose rungs.
- Inspect for corrosion or damage.
- If using a fiberglass ladder, ensure the fiberglass has not deteriorated.
- If the ladder has safety shoes ensure they are secure and in good shape.
- If working around electricity, remember that a metal ladder is a conductor.
- Only one person at a time on the ladder.
- Keep your hands free for climbing the ladder. Do not have tools in your hand(s).
- If reaching up, do not let your waist extend beyond the top rung.

For straight or extension ladders following these rules:

- Use the four-to-one rule, meaning put the base of the ladder one foot away from the wall for every four feet of height. As an example, if the support point is 16 feet above the ground put the base of the ladder four feet away from the wall.
- The ladder should extend three feet beyond the top of the support point.
- Tie the ladder off at the top with rope to anchor it. If the ladder cannot be tied off, have someone on the ground support it.

One way sites protect their workers who must often climb towers or tanks, is by the vessels having caged ladders and railings (see Figure 13-6). In the case of field tanks, if the tanks are close enough bridges may connect the tanks so that the employee doesn't have to climb each tank individually.

© Cengage Learning 2013

Figure 13-6 Caged Ladder and Tank Bridges

FALL PROTECTION

Process technicians will be required to work at a variety of heights, such as near the top of a 150 foot tall distillation tower or on the top of a small 10 foot high storage tank. A large number of deaths occur from a fall in the eight foot range. Because of this, OSHA mandated that workers need fall protection if working at an elevation of six feet or greater unless proper protection already exists, such as railings or cages. Fatal falls were 617 in 2009 versus 700 in 2008, a decline of seven percent (Bureau of Labor Statistics, 2010). About half of all the fatalities occurred in construction.

A fall protection program requires the identification of the potential fall hazards in a site workplace. Any time a worker is at a height of six feet or more the worker is at risk and needs to be protected. The two ways of accomplishing this are engineering controls and fall protection equipment. Engineering controls can be as simple as moving the work to ground level and eliminating the work height or the addition of platforms, railings, and toe boards to provide permanent, secure access to high maintenance areas and devices. When engineering controls are not feasible or practical, such as at construction or maintenance projects, a personal fall protection system is employed to prevent injuries from falls.

Fall Protection Systems

Fall protection systems can consist of devices that arrest a free fall or devices that restrain a worker in position to prevent a fall from occurring. To arrest a fall in a controlled manner, it is essential that there is sufficient energy absorption capacity in the system. Without this designed energy absorption, the fall can only be arrested by applying large forces to the worker and to the anchorage, which can result in the worker being injured even though their fall was stopped or the anchorage failing and the worker consequently falling.

A **fall arrest system** is employed when a worker is at risk of falling from an elevated position. A positioning system restrains the elevated worker, preventing them from getting into a hazardous position where a fall could occur, and also allows hands-free work. Both systems have three components: harnesses or belts, connection devices, and tie-off points.

Harnesses and Belts

Full body harnesses (see Figure 13-7a) wraps around the waist, shoulders, and legs. A D-ring (see Figure 13-7b) located in the center of the back provides a connecting point for lanyards or other fall arrest connection devices. In the event of a fall, a full-body harness distributes the force of the impact throughout the trunk of the body—not just in the abdominal area. This allows the pelvis and shoulders to help absorb the shock, reducing the impact and possible harm to the abdominal area. Maximum force arrest on a full-body harness, which is used for the most severe free fall hazards, is 1,800 pounds. Full-body harnesses come with optional side, front and shoulder D-rings. The side and front D-rings are connection points used for work positioning, and the shoulder D-rings are for retrieval from confined spaces. Three factors determine the arresting force from a fall: lanyard material type, free fall distance, and the weight of the worker. The use of a shock-absorbing lanyard or a higher tie-off point will reduce the impact force.

Belts are used in positioning system applications. Belts have two side D-rings, and are used only for restraining a worker in position. This type of belt is not used for any vertical free fall protection.

Figure 13-7a   Harness and Lanyard

Figure 13-7b   D-Ring on Harness

Connection Devices

Connection devices attach the belt or harness to the final tie-off point. This can be one device, such as a lanyard, or a combination of devices, such as lanyards, lifelines, worklines, rope grabs, and tie-off straps. Lanyards are used both to restrain workers in position, and to arrest falls. When using a lanyard as a restraining device, the length is kept as short as possible. A restraining lanyard should not allow a worker to fall more than two feet. Restraining lanyards are available in a variety of materials, including steel cables, rebar chain assemblies, and nylon rope. Fall protection lanyards can be made of steel, nylon rope, or nylon or Dacron webbing. Fall protection lanyards may also have a shock-absorbing feature built in, thus reducing the potential fall arrest force. Remember that maximum arrest

176

force is 900 pounds for belts, or 1,800 pounds for full-body harnesses. With a belt, the use of a shock-absorbing lanyard is recommended because it limits the arresting force from a six-foot drop to 830 pounds. If a shock-absorbing lanyard is not used, the tie-off point must be high enough to limit the arrest force to less than the 900-pound limit. The height of this tie-off point will vary, depending on the lanyard material and the weight of the person involved. A lanyard used for a fall is limited to allow a maximum six-foot free fall. For this reason, most lanyards are a maximum of six feet long. However, if a higher tie-off point is used, the lanyard can be longer if the free fall distance does not exceed six feet.

Lifelines

The most common fall arrest system is the vertical lifeline: a stranded rope that is connected to an anchor above, and to which the user's PPE is attached either directly or through a "shock absorbing" lanyard. Once all of the components of the particular lifeline system meet the requirements of the standard, the anchor connection is then referred to as an anchorage, and the system as well as the rope is then called a "lifeline." Lifelines add versatility to the fall arrest system. When used in conjunction with rope grabs, a lifeline allows the worker to move along the length of the line rather than having to disconnect and find a new tie-off point. The rope grab is engineered to arrest a fall instantly. A rope grab and lifeline system is a passive form of protection, allowing the user to move as long as tension is slack on the lifeline. If a fall occurs, the tension on the rope grab triggers the internal mechanism to arrest the fall. Retractable lifelines automatically retract any slack line between the worker and the tie-off point. While this type of line doesn't require a rope grab, it must be kept directly above the worker to eliminate any potential swing hazard if the worker falls.

Tie-Off Points

A **tie-off point** is where the lanyard or lifeline is attached to a structural support. This support must have a 5,000-pound capacity for each worker tying off. Workers must always tie off at or above the D-ring point of the belt or harness. This ensures that the free fall is minimized, and that the lanyard doesn't interfere with personal movement. Workers must also tie off in a manner that ensures no lower level will be struck during a fall. To do this, add the height of the worker, the lanyard length, and an elongation factor of 3.5 feet. Using this formula, a six-foot tall worker requires a tie-off point at least 15.5 feet above the next lower level.

Other Devices

For confined space applications, a tripod and winch system is used as both the tie-off point and connection device. It is used in conjunction with a full-body harness to lower and raise workers into tanks or manholes. Never use a material-handling device for personnel unless it is specifically designed to do so.

Inspection and Maintenance

OSHA regulations require that all fall arrest equipment be inspected prior to its use. This includes looking for frays or broken strands in lanyards, belts, and lifelines, and oxidation or distortion of any metal connection devices. To properly maintain the devices, periodic cleaning is necessary. Clean all surfaces with a mild detergent soap, and always let the equipment air dry away from excess heat. Follow the manufacturer's instructions for cleaning and maintenance. *Because of the stress equipment sustains from a fall, any equipment exposed to a fall must be taken out of service and not used again for fall protection.*

SUMMARY

Personal protective equipment (PPE) is last in the hierarchy of hazard control. PPE is less desirable than engineering and administrative controls for the control of hazards, but is still critical. The worker must be knowledgeable about the limitations of each piece of PPE. Respirators and gloves used improperly or chosen in ignorance can lead to exposures nearly as high as that of an unprotected worker. Equipment to protect the body against contact with known or anticipated toxic chemicals has been divided into four categories according to the degree of protection afforded. Those categories are:

- Level A should be worn when the highest level of respiratory, skin, and eye protection is needed.
- Level B should be worn when the highest level of respiratory protection is needed, but a lesser level of skin protection.
- Level C should be worn when the criteria for using air-purifying respirators are met.
- Level D should be worn only as a work uniform and not on any site with respiratory or skin hazards. It provides no protection against chemical hazards.

There is no universal chemically resistant suit that is safe to use with all chemicals. PPE selection is a process that involves a mix of experience, common sense and quantitative risk assessment.

A fall protection program requires the identification of the potential fall hazards in a site workplace. Any time a worker is at a height of six feet or more, the worker is at risk and needs to be protected. The two ways of accomplishing this are engineering controls and fall protection equipment. Fall protection systems can consist of devices that arrest a free fall or devices that restrain a worker in position to prevent a fall from occurring. To arrest a fall in a controlled manner, it is essential that there is sufficient energy absorption capacity in the system. Without this designed energy absorption, the fall can only be arrested by applying large forces to the worker and to the anchorage.

REVIEW QUESTIONS

1. (T/F) Personal protective equipment (PPE) is last in the hierarchy of hazard control.

2. The function of _____ is to protect the user's entire body, including the respiratory system, eyes, hearing, head, hands, etc.

3. List five types of PPE.

4. _____ creates a protective barrier between the worker and the hazard.

5. _____ protection should be worn when the criteria for using air-purifying respirators are met.

6. _____ protection should be worn when the highest level of respiratory, skin, and eye protection is needed.

7. Assume you and a coworker can't decide what level to don in a particular situation. In that case, you should don _____ .

8. _____ is the amount of chemical that will pass through a material in a given area in a given time.
   a. Penetration
   b. Degradation
   c. Permeation

9. _____ is the gradual chemical destruction of a material.
   a. Penetration
   b. Degradation
   c. Permeation

10. By _____ it is meant how long it takes for the chemical to permeate through the glove material and make contact with the skin.

11. (Complete this sentence) Hard hats should be stored out of direct sunlight because _____ .

12. (T/F) Safety shoes must be worn if an employee is frequently (at least once/week) exposed to potential foot hazards.

13. (Circle one) A worker (should or should not) wear flammable clothing over their FRC.

14. List four hazards that eye protection protect the worker from.

15. _____ form a seal around the eyes to provide complete protection from particles and dust.

16. A _____ is employed when a worker is at risk of falling from an elevated position.

17. A _____ restrains the elevated worker, preventing them from getting into a hazardous position where a fall could occur, and also allows hands-free work.

18. A fall arrest or positioning system as the following three components: _____ , _____ , and _____ .

19. Maximum force arrest on a full-body harness, which is used for the most severe free fall hazards, is _____ pounds.

20. Two common connection devices that attach the belt or harness to the final tie-off point can be _____ or _____ .

21. A _____ is where the lanyard or lifeline is attached to a structural support. This support must have a _____ capacity for each worker tying off.

22. (True or false) Fall protection equipment that has experienced a fall may be reused.

EXERCISES

1. Scenario: A gas, methyl isocyanate (MIC), is leaking from a blown flange gasket. When you step outside the control room your eyes begin to water and you begin to cough. You go back inside just as an alarm sounds in the control room and an outside sensor reports 300 PPM MIC in the ambient atmosphere outside the control room.

Instructions

Download a methyl isocyanate MSDS from the Internet and answer the following questions:

1. What level of PPE will you don to search for and contain the leak?
2. List each item of PPE you will wear.

2. Research on the Internet (1) the level of impact for which safety glasses are designed, and (2) the level of impact for which hard hats are designed to withstand. List the sites from which you obtained the information.

RESOURCES

www.wikipedia.org

www.osha.gov

www.YouTube.com

www.cdc.gov/niosh/

www.toolboxtopics.com

www.ehow.com

CHAPTER 14

Hazard Communication (HAZCOM)

Learning Objectives

Upon completion of this chapter the student should be able to:

- *Describe the six critical items required in the Hazard Communication program.*

- *Explain the purpose of the Hazard Communication program.*

- *Describe four information requirements of the HAZCOM standard.*

- *Explain what must be on the label of each hazardous chemical manufactured.*

- *Explain the importance of the material safety data sheet.*

INTRODUCTION

After almost 10 years of aggressive enforcement, Occupational Safety and Health Administration's (OSHA) Hazard Communication rule remains the agency's standard that results in the most numerous fines and penalties. From October 2001 to September 2002, OSHA wrote 2,073 citations for Hazard Communication (HAZCOM) violations that totaled $680,000. The citations occurred because most of the hazard communication programs and their site enforcement were very deficient. In 2010, Hazard Communication violations remained high on the list of the top 10 violated standards and unchanged at number three on the list (Thomas Galassi, 2010, MSDSonline Environmental, Health & Safety Blog).

The Hazard Communication Standard (HAZCOM) is *performance-based*, not *specification-based*. Performance-based standards are very flexible. They tell you what you have to accomplish and you decide how to do it. Except for a few minor things like ensuring that shipped containers have labels in English, OSHA doesn't care how you meet the goals of the hazard communication standard as long as you meet them.

OSHA HAZARD COMMUNICATION STANDARD

A written hazard communication program is a blueprint which describes how a company will achieve the goals set by the Hazard Communication standard. The program must address six critical items:

- Material safety data sheets
- Container labeling
- Hazards in unlabeled pipes
- Non-routine tasks
- Information exchange with other employers
- Employee training and information

Under authority of the Occupational Health and Safety Act in 1994, OSHA issued its **HAZCOM Standard** to address the assessment and communication of chemical hazards in the workplace. OSHA has authority to enforce its Hazard Communication standard (see Figure 14-1). Civil penalties (fines) and criminal penalties (for gross neglect) are possible

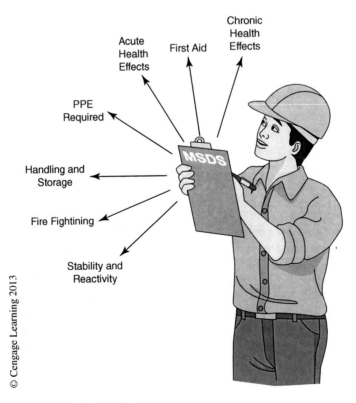

Figure 14-1 Workers Right to Know

© Cengage Learning 2013

for violations of this regulation. Civil penalties for any violation of OSHA regulations can range up to $10,000. This standard has often been referred to as "the worker's right to know" standard. This standard is intended to help protect employee safety and health in work sites where chemicals are used or are present. Its purposes are to ensure that:

- The hazards of chemicals which are produced or imported are evaluated for health and safety hazards.
- Employees who come into contact with these chemicals in their workplace are informed about the hazards and how to protect themselves from the hazards.
- The necessary information is conveyed comprehensively through a Hazard Communication Program.
- The labeling system in use is clearly explained.
- Information contained in a material safety data sheet (MSDS) is clearly explained.
- Training or information is presented about detecting the presence or release of a hazardous chemical.
- Training or information is presented about how to respond in the event of an emergency involving the chemical.

Where only sealed containers of chemicals are handled, as in warehousing or retail sales, employees only have to be trained as necessary to protect themselves in the event of a spill or leak. No written hazard communication program is required.

There is no list of hazardous chemicals. Instead, OSHA has defined two categories of hazardous materials: **health hazards** and **physical hazards**. If a material meets one of OSHA's definitions it is considered a hazardous chemical.

- *Health Hazards* are defined as materials for which there is scientific evidence demonstrating that acute or chronic health effects may occur in exposed employees. Health hazards include carcinogens, toxic agents, irritants, corrosives, sensitizers, and any agent which damages the lungs, skin, eyes, or mucous membranes:
- *Physical Hazards* are things that are hazardous because of their physical properties. They include combustible and flammable liquids, compressed gases, explosives, oxidizers, and highly reactive materials.

EMPLOYEE TRAINING

The most important element of the entire right-to-know program is employee training and education. In the processing industry all employees—operators, analyzer and instrument technicians, and maintenance personnel—are at risk from hazardous chemicals. Employers are required by the HAZCOM standard to provide employees with effective information and training on hazardous chemicals in their work area. The employee training section should contain a synopsis of the educational program. It should provide a description of how the company intends to train its employees about routine hazards and hazards created by non-routine tasks.

Training Requirements

Employees must be effectively trained when they are assigned to a new task for which training has not yet been provided or whenever a new physical or health hazard is introduced into the work area. Training means that information conveyed must be understood. It is not sufficient, for example, to give the employee an MSDS and tell them to read it. If the workers do not understand English, training must be conducted in a language they do understand. Training must include, at minimum, the physical and health hazards of chemicals in the work area.

Training must be given at the time of initial assignment and whenever a new hazard is introduced. At a minimum, the training program must include the:

- Nature of hazards posed by chemicals in the workplace
- Measures that employees can take to protect themselves from these hazards
- Instructions on work practices, personal protection equipment (PPE), and any special procedures to be followed in an emergency
- An explanation of the hazard communication program, including information on labeling and MSDSs

Information Requirements

Employers must inform employees about:

- The requirements of the OSHA HAZCOM standard
- Any operations in their work area where hazardous chemicals are present
- The location and availability of the employer's written Hazard Communication Program
- Where to find the list of hazardous chemicals used by the company
- How to access an MSDS

RESPONSIBILITY FOR HAZARD DETERMINATION

A hazard determination is an objective evaluation of the hazards which a chemical presents. It identifies and considers the available scientific evidence concerning the physical and health hazards of the chemical. Responsibilities for performing the hazard determination are that:

- Manufacturers determine hazards for the chemicals they make.
- Importers determine hazards for chemicals they import.
- Distributors determine hazards for chemicals which they mix, blend, or change the chemical composition. Distributors who just repackage do not have to do a hazard determination.

HAZARD COMMUNICATION PROGRAM (29CFR 1910.1200 E)

Every company which uses hazardous chemicals must have a written hazard communication program that must be available to employees and be provided to OSHA upon request. The Hazard Communication Program must include:

- A list of all hazardous chemicals used at the site. There can be one list for the entire facility, or, if there are different areas within the facility which handle different chemicals, there can be separate lists for the individual areas.
- The procedures for meeting requirements for labels and MSDSs.
- An explanation of how workers will be informed and trained about the hazards of chemicals to which they may be exposed.

- An explanation of how workers will be informed about the hazards of non-routine tasks (such as cleaning reactors) and the hazards of chemicals in unlabeled pipes.
- If there are other employers on the site, such as contractors, the HAZCOM Program must also state: (1) how these other employers will be informed of the site labeling system; (2) how they should protect their own employees during normal operating conditions and emergencies; and (3) how contractors and contractor-type personnel will be provided access to MSDS.

Labels

Labels on containers of hazardous chemicals provide immediate information about the contents of the container and the hazards associated with the contents. Labels are important for use in determining handling methods and immediate emergency response action. All containers of hazardous chemicals must be appropriately labeled to show the hazards of the contents and other necessary information. The information on labels must be kept current and labels must be revised within three months whenever significant new information is known about the hazards of the chemical. Labels must be printed in English. They may be in other languages as well, but never only in another language. Labels must be legible and prominently displayed on the container. The HAZCOM standard does not require a specific type of label to be used. However, labels used for commercial shipments must be consistent with requirements of the DOT Hazardous Materials Regulations (49CFR).

Chemical manufacturers, importers and distributors must ensure that each container of hazardous chemicals leaving their workplace is labeled, tagged, or marked with the following information:

- The identity of the hazardous chemical displayed on the hazard warning label
- The name and address of the chemical manufacturer, importer, or other responsible party

Employers must ensure that each container of hazardous chemical in their workplace is labeled, tagged, or marked to identify the chemical and its specific hazards. Hazard warning labels must remain on the container even when empty until the container has been cleaned and purged. There are exceptions to the labeling requirements just discussed. For instance, labels are not required for:

- Process equipment if there is some other method (signs, operating procedures, batch tickets, etc.) that convey the required information to workers.
- Piping.
- Portable containers which are used only by the person who transferred the material into the container. However, if the portable container could be used by more than one person, or it might be left unattended for even a short time, it must be labeled the same as any other container of that chemical.

NFPA AND HMIS WARNING LABELS

There are several forms of warning labels available that provide information about the hazards of chemicals in the workplace. Two of the most commonly used are the Hazardous Material Identification System (HMIS) warning label and the National Fire

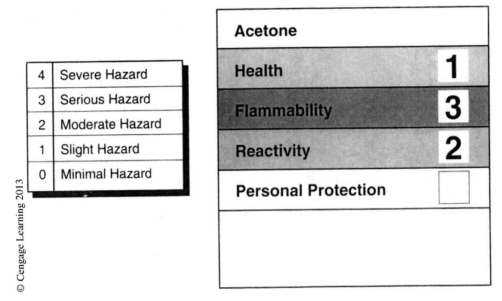

4	Severe Hazard
3	Serious Hazard
2	Moderate Hazard
1	Slight Hazard
0	Minimal Hazard

Acetone

Health	**1**
Flammability	**3**
Reactivity	**2**
Personal Protection	

© Cengage Learning 2013

Figure 14-2   HMIS Labeling System

Protection Association (NFPA) warning label. These labels are applied to vessels and other containers to identify specific hazards—flammability, health, reactivity, etc.—associated with the chemical contents.

The **HMIS label** (see Figure 14-2) is a rectangular label with horizontal color bars indicating the type of hazard (red for flammability, yellow for reactivity, and blue for health hazard), and numbers showing the degree or severity of each hazard as follows:

 0 = minimal hazard
 1 = slight hazard
 2 = moderate hazard
 3 = serious hazard or high hazard
 4 = severe hazard or extreme hazard

The HMIS label has a white space at the bottom of the label where PPE is specified.

Another acceptable label is the NFPA's diamond label (see Figure 14-3), which uses the same color code as the HMIS label, but has different and more specific meanings for the numbers in each hazard category. On this label, the number in the red diamond indicates flash point or combustibility and the numbers stand for:

 0 = will not burn when exposed to a temperature of 1,500°F for 5 minutes
 1 = flash point above 200°F
 2 = flash point between 100°F and 200°F (materials that must be moderately heated before ignition will occur)
 3 = flash point between 73°F and 100°F (easily ignitable)
 4 = flash point below 73°F

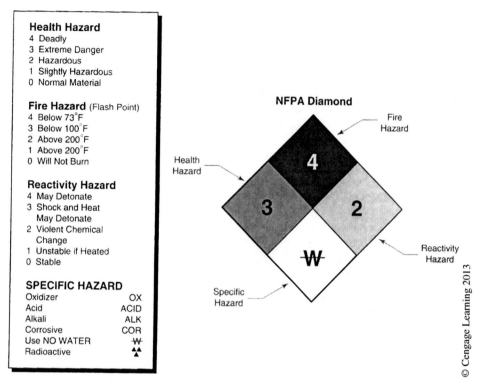

Health Hazard
4 Deadly
3 Extreme Danger
2 Hazardous
1 Slightly Hazardous
0 Normal Material

Fire Hazard (Flash Point)
4 Below 73°F
3 Below 100°F
2 Above 200°F
1 Above 200°F
0 Will Not Burn

Reactivity Hazard
4 May Detonate
3 Shock and Heat
   May Detonate
2 Violent Chemical
   Change
1 Unstable if Heated
0 Stable

SPECIFIC HAZARD
Oxidizer   OX
Acid   ACID
Alkali   ALK
Corrosive   COR
Use NO WATER   W̶
Radioactive   ☢

NFPA Diamond

Fire Hazard

Health Hazard

Reactivity Hazard

Specific Hazard

© Cengage Learning 2013

Figure 14-3   NFPA Labeling System

The number in the blue diamond represents the health hazard of the material.

0 = No health hazard under fire conditions other than for normal combustible material.

1 = Slight hazard. Exposure will cause irritation but only minor residual injury, even if prompt treatment is not given.

2 = Hazardous. Intense or continued exposure could cause temporary incapacitation or residual injury without prompt medical treatment.

3 = Highly hazardous. Short term exposure could cause serious injury, even with prompt medical treatment.

4 = Deadly. Very short exposure could cause death or major residual injury, even with prompt medical treatment.

The number in the yellow diamond indicates reactivity.

0 = Stable

1 = Normally stable, but can become unstable if heated

2 = Normally unstable and readily undergoes chemical change, but does not detonate

3 = Can be detonated, but not easily, or may react explosively with water

4 = Can readily be detonated or can decompose explosively

The lower white diamond indicates specific hazards or information, such as oxidizer, acid, alkali, or radiation hazard, or may have a *no water* symbol, which means use no water in the event of fire.

MATERIAL SAFETY DATA SHEETS

MSDSs are perhaps the most important documents in a HAZCOM program. They provide specific, detailed information about the hazards of chemicals and immediate action to take in the event of exposure to the chemical. All manufacturers and importers must obtain or develop a MSDS for each hazardous chemical they produce or import. A distributor who blends or mixes chemicals is considered to be a manufacturer.

The MSDS must be available to employers, printed in English (although other languages can be used as well), and current. If any new, significant information becomes known about the chemical's hazards, the MSDS must be revised within three months. MSDS must be sent to customers either with the initial shipment of the product or before the shipment.

At a minimum the material data sheet section of a HAZCOM program must answer the following questions:

- Who is responsible for obtaining the MSDS?
- Who will maintain the MSDS file?
- How will the MSDS file be kept up-to-date?
- What action will be taken if a data sheet is not received with an initial shipment of hazardous materials?
- How do employees gain access to an MSDS?

The author of this text considered HAZCOM as the process employee's best friend because of its requirement for a MSDS which is accessible in all the control rooms either as documents in binders or electronic documents on the site intranet. As a supervisor, the author advised all newly hired process employees to read the MSDS for each of the chemicals on their unit and take careful note of all NFPA hazard ratings of three or four. They were to think of a hazard rating of three or four as a coiled cobra prepared to strike. If the cobra bites them, it will hurt them very badly if not kill them. If they make a mistake with a chemical that has a hazard rating of three or four it is going to hurt them badly, if not kill them.

Sections of an MSDS

The MSDS must show the identity of the chemical on the label. For a single substance, the MSDS identifies its chemical and common name. If the chemical is a mixture, the MSDS must show the common name of the mixture and the chemical name or common names of ingredients which contribute to known hazards. Because MSDSs are so important to the HAZCOM process, the HAZCOM standard establishes detailed requirements for information that is needed on them. To illustrate this information, a commonly used MSDS format, the American National Standards Institute (ANSI) format (see Table 14-1) will be described. This format was developed by the Chemical Manufacturer's Association (CMA) and approved by ANSI.

The ANSI format MSDS provides all the information required to be included on an MSDS. A discussion of this information, as it appears in the different sections of this MSDS, follows. Other formats may be used, provided they convey the necessary information.

Table 14-1 MSDS Sections (ANSI)

MSDS Sections (ANSI)
1. Product name and company identification
2. Product composition
3. Hazards identification
4. First Aid
5. Fire Fighting Measures
6. Accidental Release Measures
7. Handling and storage
8. Exposure control and personal protection
9. Physical and chemical properties
10. Stability and reactivity
11. Toxicological information
12. Environmental effects
13. Disposal information
14. Transportation information
15. Regulatory information
16. Other information

© Cengage Learning 2013

Section 1—Product and Company Identification

This section includes the product name and synonyms, proper shipping name, manufacturer's name, address, telephone number, and 24-hour emergency telephone number.

Section 2—Product Composition

This section must include a listing of all health hazard ingredients which comprise 1 percent or more of the product, all carcinogenic ingredients which comprise 0.1 percent or more of the product, any ingredient which could be released from the mixture in concentrations exceeding the OSHA PEL (permissible exposure level) or the American Conference of Governmental Industrial Hygienists (ACGIH) TLV (threshold limit value), exposure limits established for each ingredient, including: OSHA Permissible Exposure Limits, (eight-hour time-weighted average), ACGIH Threshold Limit Values (eight-hour time weighted average), 15-minute Short Term Exposure Limits (STEL), and ceiling values (concentration which must never be exceeded).

Section 3—Hazards Identification

This section describes the characteristics of the material, such as color, odor, appearance, the acute health effects, and the chronic health effects.

Section 4—First Aid

This section advises what to do in case of inhalation, eye contact, skin contact, or ingestion. If inhaled, instructions say to move victim to fresh air, but may also say to administer oxygen or artificial respiration if necessary. If eye contact is made flush eyes with water for at least 15 minutes, then contact a physician. If skin contact, it usually says to flush eyes with water for at least 15 minutes, then call a physician. If the chemical is ingested, it usually indicates whether vomiting should or should not be induced.

Section 5 — Fire Fighting Measures

This section identifies whether the product is flammable or combustible, its upper and lower explosive limits, its flash point, autoignition temperature, hazardous byproducts of combustion (such as toxic gases), and appropriate fire fighting media (water, carbon dioxide, and foam).

Section 6 — Accidental Release Measures

This section identifies what to do in case of a spill or release. Trained responders and those who first arrive on the scene use it. If there is a reportable quantity, the minimum amount which requires the spill or release to be reported to the Environmental Protection Agency (EPA), it may be listed in this section.

Section 7 — Handling and Storage

This section describes how the material should be handled and stored safely to prevent spills, fire, explosion, or reactions with other materials. While some sections in the MSDS include highly technical information, handling and storage information is usually in clear, simple language. Typical advice may include storing in a well-ventilated area; in a cool, dry place; out of sunlight; and away from flammable materials.

Section 8 — Exposure Control and Personal Protection

This section tells how to prevent injury and illness due to chemical exposure. It includes eye protection, respiratory protection, protective clothing, engineering controls to prevent exposure, such as exhaust ventilation or dust collection, and may include exposure limits and advice about monitoring workers.

Section 9 — Physical and Chemical Properties

This section includes the physical state of product (solid, liquid, and gas), its physical appearance, such as color, and physical properties, such as melting point, boiling point, solubility, and evaporation rate.

Section 10 — Stability and Reactivity

This section describes hazardous reactions that might occur, such as polymerization, what chemicals are incompatible with the product, and whether there are hazardous products of decomposition.

Section 11 — Toxicological Information

This section is often very lengthy, and describes known toxic effects of the material on animals and people. This information is highly technical, and is mainly for specialists such as physicians, toxicologists, and industrial hygienists.

Section 12 — Environmental Effects

This section describes the effects of a spill or release on the environment. It may include data on aquatic toxicity.

Section 13—Disposal Information

This section briefly describes disposal information and often has general language, such as "dispose of in accordance with applicable federal, state, and local regulations."

Section 14—Transportation Information

This section states whether the material is regulated by the U.S. Department of Transportation (DOT) as a hazardous material. If it is, information in this section will include the hazard Class, United Nations (UN) or North America (NA) identification number, required labels, and restrictions on air or sea transport.

Section 15—Regulatory Information

This section includes special regulatory information applicable at the federal, state, or local level. Typical statements might be (1) does not contain substances reportable under SARA Title III or considered hazardous as defined by OSHA, and (2) listed in TSCA Inventory.

Section 16—Other Information

This section contains any other information deemed important by the producer of the MSDS. It may include explanations of abbreviations or date of the most recent revision.

TRADE SECRETS

A chemical manufacturer, importer, or employer may withhold a specific chemical identity, including the name of the chemical and other specific information, from the material data sheet, provided that the claim of "trade secret" can be supported, the necessary information concerning the material's properties and effects is disclosed, and the MSDS says that the specific identity is being withheld as a trade secret.

SUMMARY

OSHA issued its HAZCOM standard to address the assessment and communication of chemical hazards in the workplace. This standard has often been referred to as "the worker's right to know" standard. This standard is intended to help protect employee safety and health in work sites where chemicals are used or present. There is no list of hazardous chemicals. Instead, OSHA has defined two categories of hazardous materials: *health hazards* and *physical hazards*. If a material meets one of OSHA's definitions, it is considered a hazardous chemical. Every company which uses hazardous chemicals, must have a written hazard communication program. This program must be available to employees, and be provided to OSHA upon request. The most important element of the entire right-to-know program is employee training and education.

Labels on containers of hazardous chemicals provide immediate information about the contents of the container and the hazards associated with the contents. Labels are important for use in determining handling methods and immediate emergency response action. All containers of hazardous chemicals must be appropriately labeled to show the hazards of the contents and other necessary information. The two most common labels are the NFPA and HMIS labels.

MSDSs are the most important documents in a Hazard Communication Program. They provide specific, detailed information about the hazards of chemical materials and immediate action to take in the event of incident or exposure to the chemical. All manufacturers and importers must obtain or develop a MSDS for each hazardous chemical they produce or import.

REVIEW QUESTIONS

1. OSHA's _____ standard remains the agency's standard that results in the most numerous fines and penalties.

2. List four critical items the HAZCOM program must address.

3. Another name for the HAZCOM standard is the "_____."

4. (T/F) OSHA has a list of all the hazardous chemicals that plants can consult.

5. _____ are defined as materials for which there is scientific evidence of acute or chronic health effects in exposed employees.

6. The most important element of the HAZCOM program is _____.

7. Labels on a container provide two important pieces of information about a chemical, and they are its _____ and _____.

8. The two most commonly used labels are the:
   a. NPFA
   b. HMIS
   c. DOT
   d. NFPA

9. MSDS is the abbreviation for _____.

10. (T/F) The MSDS must always be available to employees.

11. Explain the importance of the MSDS to an employee working with chemicals.

12. Acute and chronic health effects in an MSDS are found in section _____.

EXERCISES

1. Understanding an MSDS

Instructions

Search on the Internet and download an MSDS for benzene. Answer the following questions. You may need to refer to several chapters in your textbook for definitions or explanations of some terms.

1. What does NFPA stand for and what is the function (purpose) of the NFPA?

2. How does the NFPA rate benzene for health hazards and flammability?

3. What are the two physical hazards of benzene?

4. Describe the health effects of short term exposure to benzene inhalation.

5. Describe the health effects of long term exposure to benzene inhalation.

6. If a pool of benzene was on fire what media would you use to extinguish it?

7. (True or false) Benzene is a carcinogen.

8. What is a carcinogen?

9. What are the lower and upper flammability limits of benzene?

10. Explain what those limits mean.

11. What is the OSHA TWA for benzene and explain what that means.

12. A gas detector indicates ten PPM benzene vapors in the air. What type of respirator would you use based on this concentration?

13. What type of respirator would you wear if the benzene concentration in air was 1,000 PPM?

14. Referring to question #13 above, why wouldn't you don a full facepiece respirator with organic cartridge?

15. (True or false) Benzene vapors, if not trapped in a room, will rise upwards into the air and be swept away by air currents. Explain your answer.

16. (True or false) Benzene has the same weight as water and thus will mix with water. Explain your answer.

17. Locate an MSDS for sulfuric acid on the Internet and answer the same questions for sulfuric acid that were asked about benzene.

RESOURCES

www.wikipedia.org

www.osha.gov

www.YouTube.com

www.toolboxtopics.com

CHAPTER 15

Respiratory Protection

Learning Objectives

Upon completion of this chapter the student should be able to:

- *Describe the three types of respirators.*

- *List four facial features that interfere with the fit of a facepiece.*

- *Describe five topics of training under the Occupational Safety and Health Administration's (OSHA) respiratory protection program.*

- *Explain what is meant by* respirator protection factor.

- *Explain the disadvantages of a self-contained breathing apparatus (SCBA).*

- *Explain the disadvantages of a supplied air respirator (SAR).*

- *Explain how an air-purifying respirator works.*

- *List two factors that should be considered when planning work with respirators.*

INTRODUCTION

The air, dust, and fumes some workers breathe at work can make them sick, plus it creates a sickening bottom line for employers. Lung diseases and respiratory illnesses can range from the common cold to lung cancer, asthma, allergies, and pneumonia. The cost in 2001 for occupational lung diseases was approximately 8.5 billion dollars (Occupational Hazards, May 2002). The National Institute for Occupational Safety and Health (NIOSH) reported that 3.3 million workers used a respirator in 2001. However, barely half of them knew why they were wearing the respirator or were taught how to use it properly. Material safety data sheets (MSDSs) were used by only 57 percent of the workplaces to determine the correct type of respirator required. Private industry employers reported 14,800 cases of occupational lung disease in 2008, while state and local government reported an additional 7,800 cases (www.lungusa.com, American Lung Association State of Lung Disease in Diverse Communities 2010).

Several types of respirators can be used to protect workers from respiratory hazards and each type is made for use in a specific environment. Not all respiratory equipment marketed is approved. Federal regulations require the use of respirators that have been tested and approved by the Mine Safety and Health Administration (MSHA) and NIOSH. Approval numbers are clearly written on all approved respiratory equipment. Management determines that respirators are mandated in certain situations and must be diligent in enforcing compliance. They should not have a policy that requires respirators but leaves compliance up to the employees.

RESPIRATORY PROTECTION

Respiratory protection is of primary importance since inhalation is one of the major routes of exposure to chemical toxicants. Respiratory protective devices (respirators) consist of a facepiece connected to either an air source or an air-purifying device (cartridge or filter). Respiratory protective equipment can be categorized into three types:

- **Air-purifying**—purifies contaminated ambient air by mechanical or chemical means.
- **Self-contained breathing apparatus (SCBA)**—contains an independent limited supply of air.
- **Supplied air respirator (SAR)**—uses air from a central source supplied through a hose.

Respirators with an air source are called *air-supplied respirators* and consist of two types, the (1) SCBAs which gets its air supply from a source (tank or bottle) carried by the user, and (2) the SAR, which is connected to the user by an air hose connected to a distant air supply. It is sometimes referred to as an airline respirator.

SCBAs, SARs, and air-purifying respirators also are differentiated by the type of air flow supplied to the facepiece. The air flow can have either *positive pressure* or *negative pressure*. Positive-pressure respirators maintain a positive pressure in the facepiece during both inhalation and exhalation. The two main types of positive-pressure respirators are pressure-demand and continuous. In pressure-demand respirators, a pressure regulator and an exhalation valve on the mask maintain the mask's positive pressure except during high breathing rates. If a leak develops in a pressure-demand respirator facepiece, the regulator sends a continuous-flow of clean air into the facepiece that prevents penetration by contaminated outside air. Continuous-flow respirators send a continuous stream of air into the facepiece at all times. With SCBAs, the continuous flow of air prevents infiltration by ambient air but uses up the air supply much more rapidly than with pressure-demand respirators. Only respirators operated in the positive-pressure mode are recommended for work at hazardous atmospheric sites.

Negative-pressure respirators draw air into the facepiece via the negative pressure created by user inhalation. The main disadvantage of negative-pressure respirators is that if a leak develops in the system (crack in the hose or ill-fitting facepiece), the user draws contaminated air into the facepiece during inhalation. Air usage is somewhat less and bottle supply lasts slightly longer.

Different types of facepieces are available for use with the various types of respirators. The two types are full facepieces and half mask. A full facepiece mask covers the face from the hairline to below the chin and provides eye protection. A half-mask covers the face from below the chin to over the nose but does not provide eye protection.

Personal Features and Respirators
Certain personal features of workers may jeopardize their safety while using respiratory equipment. These are determined by the work site. Some safety factors to keep in mind are:

- Facial hair located along the sealing surface of the facepiece may prevent a satisfactory seal between the mask and skin. Even a few days growth of facial hair will allow contaminant penetration inside the respirator. For this reason, beards are prohibited in petrochemical and refining sites.
- Eyeglasses with conventional temple pieces (earpieces) interfere with the respirator-to-face seal of a full facepiece. A spectacle kit should be installed in the facemasks of workers that require vision correction.
- Contact lenses may trap contaminants and/or particulates between the lens and the eye and cause irritation, damage, and an urge to remove the respirator. Wearing contact lenses with a respirator in a contaminated atmosphere is prohibited by the Occupational Safety and Health Administration (OSHA).

Respirator Fit Testing
The fit or integrity of the facepiece-to-face seal of a respirator affects its performance. A secure fit is important with positive-pressure equipment and is essential to the safe functioning of negative-pressure equipment. Most facepieces fit only a certain percentage of the population. Because of this, each facepiece must be tested on the potential wearer to ensure a tight seal. Facial features such as scars, hollow temples, prominent cheekbones, deep skin creases, dentures, or missing teeth may interfere with the respirator-to-face seal. Many companies hire outside contractors to do their respirator fit testing. The contractors use respirators integrated to a computer that has a special software program that monitors for a leaking facepiece during fit testing. The computer prints out the test result which is be kept on file for regulatory compliance.

Respirator Training
Respirator training is required by OSHA under CFR 29 Part 1910.134. The standard basically states that in any workplace where respirators are necessary to protect the health

of the worker the employer will develop and implement a respirator protection program that has a training activity. Some of the training under the program includes:

- Procedures for selecting respirators
- Medical evaluations of employees required to use respirators
- Fit-testing procedures
- Procedures for proper use of respirators in routine and emergency situations
- Procedures and schedules for cleaning, disinfecting, storing, repairing, and otherwise maintaining respirators
- Procedures for adequate air quality and flow to air-supplied respirators
- Training of employees to respirator hazards to which they are exposed during routine and emergency situations

Training should be completed prior to usage in a hazardous environment and should be repeated annually. The discomfort and inconvenience of wearing a respirator creates an unconscious resistance by some workers not to don respirators when they should. This puts them at risk. This resistance can be overcome by good training and enforcement of the company's program.

Respirator Protection Factor

The level of protection that can be provided by a respirator is indicated by the respirator's protection factor (PF). This number, determined experimentally by measuring facepiece seal and exhalation valve leakage, indicates the relative difference in concentration of the concentrations of substances outside and inside the facepiece that can be maintained by the respirator. For example, a PF of 10 for a respirator means that a user could expect to inhale no more than one tenth of the airborne contaminant present. A source of protection factors for various types of atmosphere-supplying and air-purifying respirators can be found in the American National Standards Institute (ANSI) standard ANSI Z-88.2-1980.

At sites where the identity and concentration of chemicals in air are known, a respirator should be selected with a protection factor that is sufficiently high to ensure that the wearer will not be exposed to the chemicals above the applicable limits. These limits include OSHA's Permissible Exposure Limits (PELs) and are designed to protect most workers who may be exposed to chemicals day after day throughout their working life. The OSHA PELs are legally enforceable exposure limits and are the minimum limits of protection that must be met.

TYPES OF RESPIRATORS

Earlier in the chapter we said that respirators could be categorized into three types: the SCBA, the SAR, and the air-purifying respirator. Each serves the same function, to give the user safe breathing air and to prevent hazardous chemicals from being inhaled and going directly into the bloodstream.

Self-Contained Breathing Apparatus

A SCBA (see Figure 15-1) usually consists of a facepiece connected by a hose and a regulator to an air source carried by the wearer. Mechanical regulators govern the pressure and amount of air supplied to the user and have bypass mechanisms in case of mechanical failure. Only positive-pressure SCBAs are recommended for entry into atmospheres that are immediately dangerous to life and health (IDLH). An IDLH atmosphere exposes

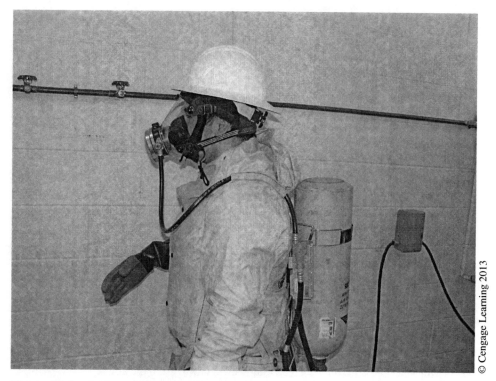

Figure 15-1   Self-Contained Breathing Apparatus

workers to airborne contaminants that are "likely to cause death or immediate or delayed permanent adverse health effects or prevent escape from such an environment." Examples include smoke or other poisonous gases at sufficiently high concentrations. SCBAs offer protection against most types and levels of airborne contaminants. The duration of the air supply is an important planning factor in SCBA use. The amount of air carried and its rate of consumption limit air supply duration. Also, SCBAs are bulky and heavy and increase the heat stress and may impair movement in confined spaces. Generally, only workers handling hazardous materials or operating in contaminated zones require SCBAs.

Supplied-Air Respirators

Supplied-air respirators (airline respirators) supply air to a facepiece via a supply line (hose) from a stationary source. SARs are available in positive-pressure and negative-pressure modes. Pressure-demand SARs with escape provisions provide the highest level of protection and are the only SARs recommended for use at hazardous waste sites. SARs are not recommended for entry into IDLH atmospheres unless the apparatus is equipped with an escape SCBA. The air source for supplied-air respirators may be compressed air cylinders or a compressor that purifies and delivers ambient air to the facepiece. All SAR couplings must be incompatible with the outlets of other gas systems used on site to prevent a worker from connecting to the wrong gas source (nitrogen, hydrogen, etc.).

SARs enable longer work periods than SCBAs and are less bulky. However, the airline impairs worker mobility and requires workers to retrace their steps when leaving an area. Also, the airline is vulnerable to puncture from rough or sharp surfaces, chemical permeation, and obstruction from falling material. For this reason SARs users carry a five minute escape bottle of compressed breathing air with them. Should the air hose become punctured they

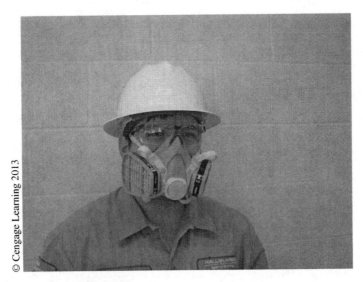

© Cengage Learning 2013

Figure 15-2   Half-Mask Respirator

can disconnect from it and exit the area breathing the escape bottle air. When in use, airlines should be kept as short as possible. Its best use is in situations where the worker must stay in a very limited space for a long time. The longest approved hose length for SARs is 300 feet (91 meters). Workers and vehicles should be kept away from the airline.

Air-Purifying Respirators

Air-purifying respirators (see Figure 15-2) consist of a facepiece with an attached air-purifying device (cartridge or canister). Air-purifying respirators selectively remove specific airborne contaminants (particulates, gases, fumes, etc.) from ambient air by filtration, absorption, adsorption, or chemical reactions. They are of two types: (1) mechanical filter respirators which offer protection only against dusts, fumes, mists, or smokes, and (2) chemical cartridge respirators which provide limited protection against organic gases. Cartridges can also provide protection against some acid gases and ammonia. They are approved for use in atmospheres containing specific chemicals up to designated concentrations. If chemical concentrations are too high, they will overwhelm the cartridge and the hazardous chemical will breakthrough. They are not approved for IDLH atmospheres.

Cartridges attach directly to the respirator facepiece. The larger-volume canisters (see Figure 15-3) attach to the chin of the facepiece or are carried with a harness and attached to the facepiece by a breathing tube. A canister serves functions similar to a chemical cartridge respirator, mechanical filter respirator, or both. It is large because it may serve multiple functions or a longer period of time. Air enters through the bottom of the canister and passes through the layers of sorbents.

Combination canisters and cartridges contain layers of different sorbent materials and remove multiple chemicals or multiple classes of chemicals from the ambient air. Though approved against more than one substance, these canisters and cartridges are tested independently against single substances. However, the effectiveness of canisters against two or more substances has been demonstrated. Filters may also be combined with cartridges to provide additional protection against particulates but this increases breathing effort and causes workers to tire quicker. Cartridges and canisters are commercially available and are color-coded to indicate the general chemicals or classes of chemicals which they protect against.

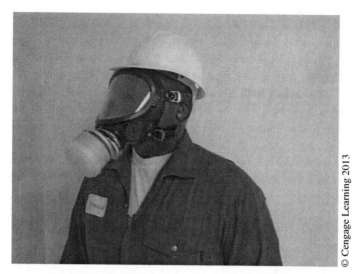

© Cengage Learning 2013

Figure 15-3   Full Face Respirator with Canister

Most chemical sorbent canisters and cartridges have an expiration date and may be used up to that date as long as they have not been opened previously. Once opened, they absorb humidity and air contaminants even if they are not in use. Cartridges should be discarded after use and should not be used for longer than one shift or when breakthrough occurs, whichever comes first.

A canister or cartridge used against gases or vapors should have adequate warning properties. NIOSH considers a gas or vapor to have adequate warning properties when its odor, taste, or irritant effects are detectable and persistent at concentrations below the recommended exposure limit. A substance is considered to have poor warning properties when its odor or irritation threshold is above the applicable exposure limit. Warning properties are essential to safe use of air-purifying respirators since they allow detection of contaminant breakthrough. Without a warning property, a worker would continue to work in a hazardous atmosphere after the sorbent ceased to function and unknowingly be breathing harmful gases or vapors. Warning properties are not fool-proof because they rely on human senses which vary widely among individuals and even in the same individual under varying conditions (head cold or allergies). Still, they provide some indication of possible sorbent exhaustion or poor facepiece fit. Air-purifying respirators can be used only when the ambient atmosphere contains sufficient oxygen (a minimum of 19.5 percent). Table 15-1 lists conditions that may exclude the use of air-purifying respirators.

Table 15-1   Conditions That May Prevent the Use of Air-Purifying Respirators

1. Oxygen deficiency
2. IDLH concentrations
3. Potential presence of unidentified contaminants
4. High relative humidity which may reduce the protection offered by the sorbent.

© Cengage Learning 2013

SUMMARY

The air, dust, and fumes some workers breathe at work can make them sick. Respiratory protection is of primary importance since inhalation is one of the major routes of exposure to chemical toxicants. OSHA requires respirator training under CFR 29 CFR Part 1910.134. The standard basically states that in any workplace where respirators are necessary to protect the health of the worker the employer will develop and implement a respirator protection program.

Respiratory protective devices (respirators) consist of a facepiece connected to either an air source or an air-purifying device (cartridge or filter). Respiratory protective equipment is of three major types:

- Air-purifiing—purifies contaminated ambient air by mechanical or chemical means.
- SCBA—contains an independent limited supply of air.
- SAR—uses air from a central source supplied through a hose.

Positive-pressure respirators maintain a positive pressure in the facepiece during both inhalation and exhalation. Negative-pressure respirators draw air into the facepiece via the negative pressure created by user inhalation. Different types of facepieces are available for use with the various types of respirators.

REVIEW QUESTIONS

1. In 2001, the cost for occupational lung diseases was about _____.

2. (T/F) All respirator equipment for sale is approved.

3. (Circle all that apply) Respirator protection is of primary importance because:
   a. It protects from flammable hazards
   b. It is important for protecting the eyes from chemicals
   c. Inhalation is a major route of exposure

4. List the three categories of respirators.

5. The main disadvantage of _____ regulators is that if a leak develops in the system the user draws in contaminated air.

6. Only respirators in the _____ mode are recommended for work at hazardous atmospheric sites.

7. Three facial features that may interfere with the seal of a respirator are:
   a. Scars
   b. Tattoos
   c. Hollow temples

d. Prominent brows

e. Dentures or missing teeth

8. List four topics of training under OSHA's respiratory protection program.

9. Respirator training is required every _____ year(s).

10. OSHA _____ are legally enforceable exposure limits for atmospheric hazards.

11. The abbreviation IDLH stands for _____.

12. Two disadvantages of a SCBA are:
    a. They require extensive training
    b. They cannot be worn by small or frail personnel
    c. They are bulky and increase heat stress
    d. They have a limited air supply

13. (True/False) All SAR coupling must be incompatible with the outlets of other gas systems used on site, including plant air.

14. Describe two disadvantages of SAR respirators.

15. (Circle those apply) The two types of air-purifying respirators are:
    a. Gas filters that protect against all gases
    b. Mechanical filters that remove dusts and smokes
    c. Airborne contaminant filters
    d. Chemical cartridge filters that provide limited protection against organic chemicals

16. (Circle one) (NIOSH, OSHA) considers a gas to have adequate warning properties when its odor, taste, or irritant effects are detectable and persist at concentrations below the recommended exposure level.

17. Air-purifying respirators can only be used when the ambient atmosphere contains a minimum of _____ percent oxygen.

EXERCISES

1. Research on the Internet the term "immediately dangerous to life and health" and write an explanation of what that term means. Next, find the concentrations of three gases and their concentrations that are immediately dangerous to life and health.

2. Go to www.YouTube.com and view two videos on respiratory protection. Write a one page report on what you learned and list the names of the videos viewed.

RESOURCES

www.wikipedia.org

www.osha.gov

www.YouTube.com

www.cdc.gov

www.cdc.gov/niosh

www.toolboxtopics.com

CHAPTER 16

Process Safety
Management (PSM)

Learning Objectives

Upon completion of this chapter the student should be able to:

■ *Explain the purpose of the Process Safety Management (PSM) standard.*

■ *List six elements of the PSM standard.*

■ *Explain why process technicians are an important part of PSM.*

■ *Explain the function of a risk assessment.*

■ *Explain why the management of change is important.*

■ *Explain how an incident investigation can be an opportunity for improvement.*

■ *List four types of process safety information.*

INTRODUCTION

Large catastrophic accidents—(Figure 16-1) fires, explosions, chemicals releases, etc.— such as at Bhopal in 1984 and in the United States at the Phillips and Arco plants in the1990s, were the stimulus that set the Occupational Safety and Health Administration (OSHA) to seeking a standard that would drastically limit the possibility of events like these happening again. The Bhopal disaster, especially, showed the importance of maintenance and administrative controls. These disasters stimulated Congress to ask OSHA to come up with a regulation to manage process safety. The result was the Process Safety Management

Figure 16-1   Picture of Texas City disaster

Standard, 29 CFR 1910.119, which became law in 1992. It mandated that industries with highly hazardous chemicals conduct process hazards analyses and develop and implement a process safety management program.

In designing the Process Safety Management standard (PSM), OSHA looked at overall process safety from a broad system's view and identified 14 key elements that industry should address to minimize catastrophic accidents. The 14 elements cover the things that can cause a process failure and accidents. Some of the major causes of process accidents are a lack of training, lack of information about process equipment, lack of equipment inspections, poor coordination with contractors, and lack of employee participation in process planning and implementation. The program is triggered by above-threshold quantities of any of 136 chemicals. The purpose of the standard is to minimize the consequences of a catastrophic release of a toxic, flammable, reactive, or explosive chemical. The importance of this standard is that it requires safety analysis and names certain analytic techniques to use or their equivalent.

PSM has brought about many positive changes since its inception. Perhaps, the biggest impact has been the change in attitude or mind-set of the workers. Today, it is impossible to think of modifying a process or equipment without first performing a pre-startup safety check. The safety check has now extended to many other aspects, such as environmental and industrial hygiene impacts. Although, this alone can't eliminate all the accidents, it has made a significant reduction in accident rates.

This is not to say that before the PSM standard nothing was done to promote safety. There existed the American National Standards Institute (ANSI) and the American Petroleum Institute (API) industry standards, and when rigorously followed, greatly reduce chance mishaps. Since these standards were already in place, why were severe accidents still occurring? Perhaps because ANSI and API standards were voluntary, but OSHA's CFR 1910.119 is not?

THE ELEMENTS OF PSM

The 14 elements that make up the PSM standard are:

- Employee participation
- Process safety information
- Process hazards analysis
- Operating procedures
- Training
- Contractors
- Pre-startup safety reviews
- Mechanical integrity
- Non-routine work authorization (Permit system)
- Management of change
- Incident investigation
- Emergency preparedness
- Compliance audits
- Trade secrets

We will take a brief look at each of the 14 elements of PSM.

Employee Participation

Safety advocates believe that management's openness to employees and allowing employee participation strongly promotes safety. Employees are a valuable resource for conducting hazard reviews, providing data for equipment reliability, or performing a pre-startup check. Process engineers are well qualified to conduct hazard reviews, identify hazards, and conduct Hazard and Operability (HAZOPs) investigations; however, operators, instrumentation technicians, and analyzer technicians are closest to the process and possess a lot of practical insight necessary for making improvements in safety.

Process Safety Information

Just as an MSDS is a crucial document for identifying the hazardous properties of a chemical, process equipment information is equally important for conducting HAZOP or pre-startup safety checks. The information should be comprehensive and accurate. A number of items can be included, some of which are:

- Piping and instrumentation diagrams (P&IDs)
- Equipment data (pressure and temperature rating, metallurgy, etc.)
- Relief valve locations, sizes, set points, and sizing data
- Corrosion and erosion effects on process equipment
- Process monitoring tools

Process Hazard Analysis

The idea behind a process hazard analysis (PHA) is to identify hazards of the process in the design stage. If the hazards are identified during the design stage, the corrective action is cheaper than discovering the hazard later. After the design phase, a PHA is used as an organized, systematic effort to identify and analyze potential hazards associated with processing or handling hazardous chemicals. There are many techniques used for hazard analysis, some of which are HAZOP, Fault-Tree Analysis, Event Tree Analysis, Checklist, and Failure Mode and Effects Analysis (FMEA). No single technique can be effective for all situations. For instance, the FMEA technique has been found to be useful for analyzing control systems or components of a system; however, it is not very effective at analyzing process hazards. For process hazards, the technique called HAZOP is most widely used.

A comprehensive risk assessment must be conducted before the analysis methodology is selected. This assessment must evaluate the degree of risk as a result of system component failures. Although this analysis can be conducted on either a quantitative or qualitative basis, the key is to ensure that both the severity and probability (frequency) of undesirable events is considered. This helps management determine whether risk is acceptable. If unacceptable, the system must be modified in order to lower either the probability, severity, or both. For example, consider a high-pressure jacketed reactor with a highly exothermic reaction. Assume that the heat of reaction is removed by pumping a coolant through the reactor jacket. Risk assessment reveals that pump failure can produce a runaway chemical reaction and explosion—an unacceptable risk. Adding a spare pump will lower the probability of explosion because two pumps must fail for the explosion to occur.

Operating Procedures

Operating procedures describe tasks to be performed, data to be recorded, operating conditions to be maintained, and safety and health precautions to be taken. Prior to PSM, a lot of training was conducted by word of mouth. Some companies had written procedures which were operationally sound (how to do the task), but failed to address safety issues related to the task. Today, procedures are written with safety in mind. The advantage of safety-based operating procedures is that it forces the user to recognize all actions in the procedure that might have a serious safety consequence.

Employee Training

All employees, including maintenance and contract employees, involved with highly hazardous chemicals must understand all associated safety and health hazards. A trained workforce is an effective deterrent against unsafe incidents. Employees of a process must be trained in the overview of the process and its operating procedures. Training must include specific safety and health hazards, emergency operations, and safe work practices. Refresher training is required every three years.

Contractors

Employers must develop a screening process to ensure that they hire only contractors who can accomplish desired jobs without compromising process and employee safety. In the past, contractors were a weak link in the safety process. As outsourcing increased with the workforce reduction, contracting became a very crucial issue. Contract personnel must be thoroughly familiar with the site safety policy and should be trained in the key safety aspects of the process unit equipment and chemicals used by the unit. Site personnel are

© Cengage Learning 2013

Figure 16-2   Checking a Blind List

responsible for training the contracted workers about potential process hazards and the worksite's emergency action plan.

Pre-startup Safety Review

The thinking behind the pre-startup safety review is to double-check everything. If something was missed, then it can still be detected before it becomes a serious problem during startup. A safety review is required for new sites and significantly modified work sites to confirm that the construction and equipment of a process are up to design specifications. The review assures that adequate safety, operating, maintenance, and emergency procedures are in place and that process operator training has been completed. Figure 16-2 shows an operator checking a blind list to ensure all blinds are either in place or removed.

Mechanical Integrity

Written procedures must be created and maintained for the ongoing integrity of process equipment. Equipment used to process, store, or handle highly hazardous chemicals must be designed, constructed, installed, and maintained so that the risk of chemical release is minimized. For new plants, mechanical integrity may not be a crucial issue, but for older plants, equipment integrity can be a major safety problem. The key is periodic inspection of equipment which is usually done during a unit turnaround (TAR). Verifying the mechanical integrity of equipment, like a vessel for instance, may include the following checks:

- Material thickness
- Stress

- Corrosion
- Mechanical integrity

There are other equivalent tests which can show that the vessel is mechanically strong. For pumps, integrity data would involve the frequency of seal repairs, bearing repairs, nature and intensity of vibration, plugging, and corrosion problems. With the mechanical integrity tests, other data is recorded, such as records of inspections and tests, maintenance procedures, establishment of criteria for acceptable test results, and documentation of inspection results. Preventive maintenance (PM) programs are a big part of the PSM.

Non-routine Work Authorization (Work Permits)

Non-routine work in a process area must be controlled in a consistent manner. Hazards identified must be communicated to those actually performing the work and to any operating personnel whose work could be affected by the hazard. PSM suggests a periodic review of the permit procedures to ensure all safety issues have been included. Work permits, such as lockout/tagout (LOTO), confined space entry (CSE), and hotwork, require special precautions since the risk of unsafe or lethal accidents is large. These permits require a thorough documentation including names of the people involved with the permit, results of the area monitoring, contingency resources required, and other items. The idea is that the detail required for the permit, forces the permit writer to be safety conscious.

Management of Change (MOC)

To properly manage changes to process chemicals, technology, equipment, and procedures, "change" must be defined. Change encompasses all modifications to equipment, procedures, raw materials, and processing conditions other than "replacement in kind." Analysis of unsafe events in the past revealed that many of the events lacked the information about changed conditions, such as revised equipment (Figure 16-3). Often, there was no documentation to show what was changed, why it was changed, when it was changed, and who changed it. Changes must be documented and a formal system must be in place that makes deviation from standard operating procedures very structured with layers of approval required and communication of change to the affected personnel.

Incident Investigation

Any incident can be looked upon as an opportunity to improve safety. Incident investigations are performed to identify underlying causes of incidents and implement steps to prevent their recurrence. The goal is to learn from past mistakes. This element requires employers to investigate as soon as possible, but not later than 48 hours after an incident resulted, or could have resulted, in a catastrophic release of chemicals. Reports must be retained for five years. This element basically has three parts: (1) assemble an incident investigation team within 48 hours, (2) address all of the findings, and (3) correct all of the findings of the investigation.

Emergency Planning and Response

Management must determine what actions employees are to take if a release occurs. Will employees handle small or minor releases? For larger releases, will employees evacuate

© Cengage Learning 2013

Figure 16-3   Failure to Manage Change

to predetermined safe zones and allow local emergency response groups to handle the release? Although many provisions of the PSM focus on preventing unsafe incidents from occurring, it is statistically impossible to have zero accidents forever. So, the thinking is that when the accidents do happen, there should be preparations and resources in place to minimize the injury and exposure to workers and loss of property. Typically, such plans may include:

- Resources for spill containment
- Resources for fire containment
- On-site fire brigade
- On-site Emergency Medical Personnel

Compliance Audits

A trained individual or team should audit the PSM program and evaluate the system's design and effectiveness. Companies are required to conduct audits every three years to ensure that they are in compliance with the PSM regulation. These audits may be conducted by the company's corporate personnel, sister company personnel, or in-plant personnel. Prompt response to audit findings and documentation that deficiencies were corrected are required.

Trade Secrets

Companies are not required to reveal trade secrets. However, in an emergency situation, safety considerations would override the issue of trade secrets.

SUMMARY

OSHA was instructed to develop a regulation to manage process safety. The result was the Process Safety Management Standard, 29 CFR 1910.119, which became law in 1992. It mandated that industries with highly hazardous chemicals conduct process hazards analysis and develop and implement a process safety management program. The 14 elements that make up the standard and their combined synergy seek to prevent or minimize the consequences of a catastrophic release of toxic, flammable, reactive, and explosive chemicals. The 14 elements cover the things that can cause a process failure and catastrophic accidents and facilitate management's efforts to prevent these from occurring.

REVIEW QUESTIONS

1. _____ led to the creation of the Process Safety Management standard (PSM).

2. Explain the purpose of the PSM standard.

3. The PSM standard consists of _____ elements.

4. The elements of the PSM standard cover the things that _____.

5. List five of the PSM elements.

6. Three reasons employees are an important part of PSM program are because they:
   a. Conduct hazards reviews
   b. Perform pre-startup checks
   c. Validate equipment inventory
   d. Provide data for equipment reliability

7. (T/F) The idea behind a process hazards analysis is to identify the hazards of the process in the design phase.

8. The advantage of _____ operating procedures is that it forces the user to recognize all actions in the procedure that might have a serious safety consequence.

9. PSM refresher training is required every _____ years.

10. (T/F) Employers do not have to screen contractor safety records before hiring them.

11. List three checks made on the mechanical integrity of a vessel.

12. _____ require special precautions since the risk of unsafe or lethal accidents is large.

13. Explain why the management of change (MOC) is important.

14. _____ are performed to identify underlying causes of incidents and implement steps to prevent their recurrence.

15. (T/F) A trained individual or team should audit the PSM program and evaluate the system's design and effectiveness.

EXERCISES

1. Go to www.YouTube.com and view two videos on any combination of the following topics: Management of change, incident investigation, and process safety management. Write a one page report summarizing the video. List the videos viewed.

2. Research the Internet and write a one page report on how the OSHA standard for process safety management came to be, describing who dates, names, places, etc. List the resources (sites visited) from which you obtained your information.

CHAPTER 17

The Permit System

Learning Objectives

Upon completion of this chapter the student should be able to:

- *List four types of permits.*

- *Explain the purpose of the permit system.*

- *Explain the purpose of the lockout/tagout permit.*

- *Explain how to lockout of a piece of equipment.*

- *Explain how to tagout an energy source.*

- *Describe three hazards that might exist in a confined space.*

- *List three elements involved in monitoring the atmosphere of a confined space.*

- *Describe the job of the attendant (hole watch).*

INTRODUCTION

The permit system is mandated under the Process Safety Management standard, the element titled "Non-routine Work." There are different permitting systems used by the chemical processing industry. Some permits are developed by a particular plant and apply only to that plant. Each plant may have its own permit system that addresses routine work and maintenance. Some permit systems, however, are required by regulatory agencies such as the Occupational Safety and Health Administration (OSHA).

A permit system requires a special document (permit), which acts like a checklist, to be filled out. Usually, the work involves some type of hazard. Personnel involved in the hazardous work must fill out a permit and the permit must be inspected and verified as complete before work can begin. The function of the permit system is to force personnel involved in a hazardous task to take the time to review all the steps, the personal protection equipment (PPE) required, the type of hazard(s) expected, and equipment required to perform the task safely. The permit system also temporarily transfers custody of a piece of equipment. As an example, a unit has a failed pump that must be repaired. This will require the pump to be removed from the process unit and taken to maintenance repair shop for repairs. The permit system places responsibilities on the issuer of the permit (operators) and the recipient of the permit (maintenance personnel). In essence, a permit system is an extra step in the direction of safety and accident prevention. Some of the more common permits are:

- Confined spaces
- Lockout/tagout (LOTO)
- Hot work
- Cold work
- Opening/blinding
- Radiation
- Critical lifts
- Electrical

There are three permit systems very common to the petrochemical and refining industries because OSHA mandates them. They are *confined space*, *lockout/tagout*, and *hot work*. All process employees—operators, instrument technicians, analyzer technicians, and maintenance personnel—sooner or later participate in a permitting situation.

HOT WORK PERMITS (29 CFR 1910.119)

The purpose of a hot work permit is to protect personnel and equipment from explosions and fires that might occur from hot work performed in an operational area. Hot work is defined as any maintenance procedure that produces a spark, excessive heat, or requires welding or burning. Examples of work considered to be hot work include grinding, welding, internal combustion engines, soldering, dry sandblasting, etc.

Areas of Responsibility

The hot work permit has multiple layers of protection and might involve several people during the issue of a hot work permit. One person might wear two hats and assume the tasks and responsibilities of two people involved in the permitting process. The responsibilities of those involved are:

- Process technician—inspects area and ensures good housekeeping, blinds, isolates and clears equipment, vessels, tanks and piping, immobilizes power driven equipment (lockout/tagout), determines PPE required, fills out the permit, and posts it at the job site.
- Process supervisor—delegates responsibilities to the process technician and ensures that all established procedures are completed.

- Maintenance supervisor—inspects area and ensures that it is ready for safety inspector, ensures that equipment, vessels, and piping are cleared, ensures that safety equipment is located near job site, reviews procedure with person performing the work, confirms PPE required, and signs permit.
- Person performing the work—inspects the job site; gathers information from process representative and mechanical supervisor about potential hazards, special procedures, or conditions; and selects and dons appropriate safety equipment before beginning work.
- Safety permit inspector—inspects area and ensures that it is safe; performs gas test and determines oxygen level; ensures that equipment, vessels, and piping are cleared; confirms required PPE; signs permit; and sets time limit. This function is performed by process technicians at many locations.
- Standby—stays in the work area and ensures that the person performing the work is safe, wears the PPE required to perform the job, warns the person performing the work if a hazardous condition develops, and calls for help, if needed.

The hot work permit must be filled out and signed before a safety inspector is called to review it and the work area. When the system is ready, the mechanical supervisor and safety inspector will show up to inspect the area. Chemical concentrations and potential hazards are assessed. The need for a standby will be determined by the mechanical supervisor and the operator. If everything is in order, the safety inspector will sign and post the permit and the work can be started. The permit must be displayed at the work site until the hot work operation is complete. The hot work permit must indicate:

1. That fire prevention and protection measures were in place before the hot work was initiated.
2. The date(s) for which the permit is approved.
3. The location and equipment used where hot work is performed.
4. That a fire watch was posted and in place during procedure and 30 minutes after work was complete.

CONTROL OF HAZARDOUS ENERGY (LOCKOUT/TAGOUT) 29 CFR 1910.147

The most effective way to control hazardous energy is to put it under lock and key. Most facilities covered by OSHA's general industry standards must implement a **LOTO program** that includes procedures and training as stated in the generic LOTO standard 1910.147. Despite the regulations, many employees continue to be fatally or severely injured each year as the result of getting caught in machines and equipment. In fiscal year 2000, OSHA issued citations for 4,149 alleged violations of 29 CFR 1910.147. Approximately one-third of these violations were issued for the lack of an energy control procedure or program. Employers are primarily responsible for protecting employees from recognized hazards. One way employers can protect their employees is by implementing general machine guards and a LOTO program with emphasis on procedures, training, and periodic inspections.

The processing industry harnesses energy from seven basic forms: electrical, pneumatic, hydraulic, compressed gases, liquids, gravity, and spring tension. Every year, industry records severe injuries because process employees fail to disconnect electrical equipment, dissipate residual energy, and restart equipment accidentally. Year after year, most LOTO injuries can be traced to one or more of these five causes, which have been given the name "The Fatal Five:"

- Failure to stop equipment
- Failure to disconnect from power source
- Failure to dissipate residual energy
- Accidental restarting of equipment
- Failure to clear work areas before reactivation of equipment

OSHA requires employers to have a written energy isolation program and to provide training to new employees upon initial assignment and every two years thereafter. Equipment modifications and new unit start-ups require the existing workforce to have additional LOTO training. The purpose of the hazardous energy standard is to protect employees from the hazards associated with the accidental release of uncontrolled energy. The LOTO procedure is a standard designed to isolate a piece of equipment from its energy source. OSHA statistics show that six percent of all workplace fatalities are caused by the unexpected activation of machines while they are being serviced, cleaned, or otherwise maintained.

In a lockout system, a padlock is placed through a gate or hasp covering the activating mechanism of an energy source, or is applied in some other manner to prevent energy from being turned on (see Figure 17-1). The lock is often color coded to indicate the division (operations, maintenance, etc.) that applied it. Often, locks from more than one division are on a lockout hasp.

© Cengage Learning 2013

Figure 17-1 Lockout and Tagout

Figure 17-2 Tagout

A tagout system is just like a lockout system except a tag is included with the lock. Tags alone should be used only in cases where a lock is not feasible (see Figure 17-2). The tags should be sturdy, waterproof, and large enough to catch the eye. They should also have a string or wire for attaching to the equipment or device to be tagged out and a place for the person doing the tagout to sign their name and date.

Lockout/Tagout Procedure

Two types of employees are involved in LOTO. They are the *affected employee* and the *authorized employee*.

- An **affected employee** is an employee whose job requires him/her to operate or use a machine or equipment on which servicing or maintenance is being performed under lockout or tagout, or whose job requires him/her to work an area in which such servicing or maintenance is being performed.
- An **authorized employee** is a person who locks out or tags out machine or equipment in order to perform servicing or maintenance on the machine or equipment.

OSHA has established a six-step procedure for locking out a piece of equipment. All of this information involved in the steps should be recorded in a lockout logbook.

1. Preparation for shutdown of the equipment. During this phase, the type of energy being isolated must be identified and the specific hazards controlled. Authorized employees must prepare for shutdown by reviewing information such as the type and magnitude of the energy, controls, and hazards.
2. Shutting down the equipment.
3. Isolation, which involves the closing of valves, shutting down main disconnects and circuit breakers, and disconnecting pneumatic, electric, hydraulic, and compressed gas and liquid lines.
4. Application of LOTO devices to breakers and disconnect switches, valves, and energy isolating devices.
5. Control of stored energy takes place by relieving pressure, grounding cables connected, elevated equipment supported, and moving parts stopped.
6. Verification that all energy hazards have been locked out. The term *lock-tag-try* is applied when the electrically disconnected equipment is checked by attempting to start the equipment at the local start-stop switch. If the procedure has been performed correctly, the equipment will not start.

The LOTO procedure must contain clear instructions covering procedures for re-energizing machines and equipment after completion of maintenance or repairs. The equipment must be inspected to verify that components are properly fastened and jumpers or grounds, if applied, are removed. Employees performing the procedure must be safely positioned before removing the LOTO devices. Affected employees not participating in the procedure must be notified that the devices have been removed.

CONFINED SPACE STANDARD (29 CFR 1910.146)

Is there carbon monoxide, hydrogen sulfide, or too little oxygen? What is in the atmosphere of the confined space the worker is about to enter? Workers can't see or feel an improper oxygen level or toxic or flammable gas levels. How are they supposed to know what's in a confined space? Approximately 50 percent of the time a confined space is hazardous because of an oxygen deficiency. Two out of three deaths in a confined space occur to persons ill-equipped or poorly trained attempting rescue.

On April 15, 1993, the OSHA rule on Confined Space Entry (CSE) 29 CFR 1910.146 became effective. OSHA's intent was to protect workers from exposure to toxic, flammable, explosive, or asphyxiating atmospheres and also from potential engulfment (burial) from powders or other free-flowing solids. The reason OSHA devoted a lot of attention to the creation of the confined space standard is that working in a confined space (CS) is especially hazardous. A confined space makes it hard to get help to a worker in trouble, plus, it is hard for the worker to get themselves out quickly. Thus, working in confined spaces requires extra precautions, and these precautions are exercised through permit requirements for entry. There are two important definitions regarding confined spaces according to OSHA.

- First, in 29 CFR 1910.146 a **confined space** is an enclosed area which (1) is large enough to enable an employee to enter and perform assigned work, (2) has limited or restricted means for entry or exit, and (3) is not designed for continuous employee occupancy.
- The second definition is a subset of the first. A *permit-required confined space* is a confined space with one or more of the following characteristics: contains or may contain a hazardous atmosphere, contains a material with the potential for "engulfment" of an entrant (such as sawdust, sand, grain, or earth), has an internal configuration or shape such that an entrant could be trapped or asphyxiated, or contains any other recognized serious safety or health hazard.

Confined Space Hazards

Flammable and toxic hazards are the most common CS hazards. Monitoring large vessels requires care even though the vessels are opened up and have blowers circulating air through them. They should be monitored at various locations since a gas, depending on its density, may be concentrated at the top of the vessel or the bottom. Engulfment hazards are a real possibility in silos and excavations. Workers can be buried alive. The most effective way to prevent this from occurring is to issue a permit only after establishing that there is no material in the silo, all blinds and lockouts are in place, and that adequate shoring of excavations are in place.

Oxygen-enriched atmospheres are a confined space hazard. Oxygen-enriched atmospheres occur when the oxygen level rises above 23.5 percent, which creates a serious fire hazard environment. Some materials not usually considered fire hazards will burn rapidly in an oxygen-rich environment. Oxygen deficient atmospheres are another CS hazard. The use of SCBA is mandatory, especially if there is a chance of encountering or developing an oxygen deficient atmosphere. OSHA regards an atmosphere of less than 19.5 percent oxygen as oxygen deficient (see Table 17-1). A number of factors can contribute to lowering the oxygen level, some of which are poor mixing of the air in the vessel, a process leak, corrosion, or other activities like welding.

Physical hazards include piping that may trip the entrant or cause the entrant to slip and fall. There could be holes or drop-offs, electric wiring, steam lines, and vessel internals

Table 17-1 Oxygen Atmospheres

Volume % Oxygen	Symptom or Comment
23.5	Maximum OSHA permissible level
20.9	Normal air
19.5	Minimum OSHA permissible level
16.0	Impaired judgement and breathing
14.0	Faulty judgement, rapid bodily failure
6.0	Difficult breathing, death in minutes

© Cengage Learning 2013

© Cengage Learning 2013

Figure 17-3 Alarm Panel Outside of Analyzer Shelter

that present hazards. Some examples of petrochemical and refining equipment that are considered confined spaces are:

- Storage tanks
- Large pipes
- Silos
- Distillation towers
- Underground cable ducts
- Analyzer shelters

It might seem strange that the last item on the bulleted list, analyzer shelters, are considered a confined space, but there are valid reasons for this category. Analyzer shelters are typically small portable building housing analyzers and sample conditioning systems that contain tubing with a constant flow of gases and liquids in them. If the tubing leaks or breaks free, the gases or liquid are released into the shelter. An analyzer technician entering the shelter may walk into a space filled with hazardous gases or vapors. For that reason, analyzer shelters have LEL/O_2 detectors mounted in them and an alarm panel mounted on the outside of the shelter by the door to warn them of a leak (see Figure 17-3).

Ventilation of Confined Spaces

Mechanical ventilation and purging are key entry preparations for vessels. If pre-entry monitoring indicates oxygen deficiency or the presence of flammable or toxic materials,

Figure 17-4   LEL and Percent Oxygen Gas Detector

the space must not be entered. First, it must be purged with forced mechanical ventilation and/or cleaned to remove all identifiable hazards. Generally, ventilation consists of portable blowers. Obviously, the compressor and the ductwork must be checked carefully for integrity and the air should be free of oil and moisture.

Gas Testing the Confined Space

CSE is a very critical operation. Issuing a permit places total reliability on gas testing devices (see Figure 17-4). The gas detector must be accurate and reliable. All the observations and adjustments during calibrations must be recorded for regulatory compliance. Maintaining instrument calibration is critical. A trained and knowledgeable person must perform testing and calibration according to manufacturer specifications. The name of that person must be recorded along with test data and dates. It is important to be thorough and take your time issuing a permit since lives are at stake. Some things to keep in mind are:

- Keep a gas detector available as a spare.
- Ensure the detector has been properly calibrated.
- Periodically communicate the instrument readings to the workers inside the CS.

Permit-Required Confined Space

The criterion for permit-required confined space is that it contains or has a known potential to contain a hazardous atmosphere, including chemicals, sludge, or sewage. *Hazardous atmosphere* means an atmosphere that may expose employees to the risk of death, incapacitation, impairment of ability to self-rescue (escape unaided), injury, or acute illness. Examples of hazardous atmospheres are a flammable gas or vapor, airborne combustible dust at a concentration that meets or exceeds its LEL, unacceptable oxygen levels, or any atmospheric condition recognized as immediately dangerous to life or health (IDLH).

If a confined space contains a material that has the potential for engulfing an entrant, it is a permit-required confined space. **Engulfment** means the surrounding and effective capture of a person by a liquid or finely divided (flowable) solid substance that can be aspirated (breathed in) to cause death by filling or plugging the respiratory system or that can exert enough force on the body to cause death by strangulation, constriction, or crushing.

Monitoring of the atmosphere inside a confined space involves three elements: (1) properly calibrated instruments, (2) established procedures, and (3) critical evaluation of results by a qualified person. The order of testing for specific atmospheric elements is important. Testing for oxygen content must be performed first, followed by tests for flammable atmospheres, then for toxic gases or vapors. Testing is done in this order because if there is not enough oxygen present, the LEL test function will not work properly since oxygen is required for that part of the instrument's operation. It may reveal an atmosphere to be safe, when it is really explosive. Normally, some benchmark, such as ten percent of LEL, is used to indicate that a potential fire hazard exists in any area when this concentration is reached. Monitoring should be continuous during all operations with the space. Concentrations of air contaminants may increase due to leaking or percolation from the adjoining environment even though initial tests revealed them to be low or non-existent.

Confined Space Entry

Let us look at the steps required before sending two men into a drained kerosene storage tank to do an inspection. They should include:

- Test and monitor the tank interior for hazards.
- Provide a means of communication between the worker(s) in the CS and an outside hole watch so that help can be summoned quickly.
- Develop plans for rescue and transportation to medical facilities.
- Alert the medical facility at your site of a confined space entry.

The purpose of the CSE program is to systematically carry out the CSE process. The major objectives of a CSE program are to:

- Keep unauthorized people from entering the CS.
- Provide tools and instrumentation necessary to ensure that entry into a confined space is safe.

Those objectives are met by:

- A confined space permit system
- A buddy system
- A list of authorized workers
- Notification system to medical services
- Annual review of the program
- Training

Before issuing a permit for entry into the storage tank (see Figure 17-5), you must check for the presence of toxicants, flammables, and other hazards in the tank. This can be deceptively simple. Imagine you are trying to measure the LEL inside the tank. At one location, the LEL shows 1 percent. You might think that the tank has no flammability danger and a permit can be issued. Be careful because this is a very large tank and the LEL may be different at different locations. For a large vessel, you must check multiple locations (top, middle, and bottom). Also, some chemicals used in a process will influence the LEL readout or differences in LEL. Viscous chemicals or slurries require extreme caution before issuing a permit. As slurries dry up, they leave a powder behind which can slowly release toxic or flammable materials. This can keep the LEL readout at a high value for a long time.

Figure 17-5 Confined Space Gas Testing

CONFINED SPACE TEAM

Safe work in a confined space requires teamwork. The confined space entry team is made up of the entry supervisor, the attendant, the entrant, and the rescue team.

Entry Supervisor

An entry supervisor is responsible to coordinate all activities related to the CS before issuing a CSE permit. They plan the entry and develop rescue plans. They check the LEL or toxic components of the confined space and the list of all LOTO items and signatures. They also assign the hole watch and entry workers. The entry supervisor has the authority to withhold issuing a permit if unsafe conditions have been detected. Since no permit remains valid beyond the duration of a shift, they have the responsibility to initiate work stoppage as the permit nears its duration.

Entrant

Confined space training is required for authorized entrants. They must be thoroughly familiar with the space and its hazards, and they should be able to detect warning signs of over-exposure. They should be physically fit to be able to get out of the vessel on their own. This may sound contradictory to the main idea about the buddy system and rescue services. But, if you think about it, the idea is that a physically able person is less likely to get trapped in the CS than a physically challenged person.

Attendant

The attendant (also called *buddy* and *hole watch*) has the primary responsibility of monitoring the safety of the persons working in the CS. They review the permit before any entry and keep unauthorized personnel out of the area. They ensure ventilation equipment is working and tend the lifeline of the entrant. The attendant does not perform any rescue function and should not enter the CS in the event of a problem. They monitor the CS worker(s) by radio or voice and summon help, if needed. They should not perform any other work, other than monitoring the atmosphere in the CS and maintain constant communication with the workers in the CS. They should have the authority to stop work at once if unsafe conditions develop and contact rescue services, if needed.

Emergency Rescue Services

Rescue services may be provided by outside contractors specializing in that specific work. Generally, plant rescue personnel or outside contractors are selected in advance and are made aware of the hazards of the CS. They should be given all the pertinent information: location, name of the vessel, permit procedures, list of hazardous chemicals, MSDS, etc. In many instances, companies decide to provide their own in-house rescue services. These employees will have received extensive training in CS rescues.

Entry into a confined space during an emergency is sometimes unavoidable. Pre-entry testing or ventilating may not be possible in situations involving a rescue. In such instances, the atmosphere must be considered immediately dangerous to life and health. A positive pressure SCBA or airline respirator with emergency escape bottle must be used. The entrant must be equipped with a full-body harness and lifeline that can be attached to a pulley and winch or hand-operated hoist which allows the hole attendant to begin rescue. Continuous communication must be maintained with the person attempting rescue.

SUMMARY

A permit system requires a permit before certain types of work can be done. Usually, the work involves some type of hazard. The function of the permit system is to force personnel involved in a hazardous task to take the time to review all the steps, PPE, hazards, and equipment required to perform the task safely. It also is required for change of custody of equipment.

The purpose of a hot work permit is to protect personnel and equipment from explosions and fires that might occur from hot work performed in an operational area. Hot work is defined as any maintenance procedure that produces a spark, excessive heat, or requires welding or burning. The purpose of the hazardous energy standard is to protect employees from the hazards associated with the accidental release of uncontrolled energy. The lockout/tagout procedure, also referred to as LOTO, is a standard designed to isolate a piece of equipment from its energy source. Confined spaces present an array of hazards. By implementing a comprehensive planning process that encompasses atmospheric testing, hazard assessment, and protect equipment selection, many of these hazards can be addressed before entry activities begin.

REVIEW QUESTIONS

1. The permit system is mandated under the Process Management Standard, the element titled _____.

2. (T/F) A plant may have its own permit system for routine work and maintenance.

3. List four types of permits.

4. Explain the function of the permit system.

5. (T/F) The function of the hot work permit is to protect personnel and equipment from the danger of a fire or explosion when hot work is performed.

6. List five people who might be involved in a permit system.

7. _____ is defined as any maintenance procedure that produces a spark, excessive heat, or requires welding or burning.

8. Fire watches must remain in place _____ after hot work is completed.

9. Four basic energy forms that the processing industry utilizes are _____, _____, _____, and _____.

10. Explain the purpose of the lockout/tagout permit.

11. _____ is the percent of workplace fatalities due to failure to follow or have an energy isolation program.

12. (T/F) Equipment modifications and new unit start-ups require the existing workforce to have additional lockout/tagout training.

13. LOTO is a standard designed to prevent:

    a. Hazardous electrical energy from being released
    b. Hazardous potential and kinetic energy from being released
    c. Equipment from being restarted
    d. Isolate a piece of equipment from its energy source

14. _____ should be used only when a lock is not feasible.

15. Define *confined space* according to OSHA.

16. Approximately _____ of the time a confined space has an oxygen deficiency.

17. Three hazards that might exist in a confined space are:

    a. Flammable atmosphere
    b. Engulfment
    c. 21 percent oxygen level
    d. Toxic atmosphere

18. (Complete this sentence) The reason why oxygen-enriched atmospheres are a confined space hazard is because _____.

19. A _____ means an atmosphere that may expose a worker to the risk of death, incapacitation, injury, or inability to self-rescue.

20. Describe the job of the attendant (hole watch).

21. _____ means the surrounding and effective capture of a person by a liquid or flowable solid substance.

22. Describe two symptoms of a 16 percent oxygen level.

23. (T/F) The LEL function on a gas tester will not work if enough oxygen is not present.

EXERCISES

1. Go to www.YouTube.com for some excellent videos on lockout/tagout and confined space. View one of each and write a one page report of what you learned that helped you understand the textbook material better. List your sources.

2. Research the Internet for a PowerPoint presentation on lockout/tagout, confined space, and hot work. Download the presentation to a flash drive or CD and bring it to class to be presented to the class either by yourself or your instructor. Cite your source.

RESOURCES

www.wikipedia.org

www.osha.gov

www.YouTube.com

www.toolboxtopics.com

CHAPTER 18

Hazardous Waste Operations (HAZWOPER)

Learning Objectives

Upon completion of this chapter the student should be able to:

- *Describe the two options HAZWOPER gives organizations for responding to a chemical spill.*

- *Explain the two important parts of a plant's operations HAZWOPER covers.*

- *Describe the three broad categories HAZWOPER can be divided into.*

- *List the five levels of emergency responders.*

- *Describe the responsibilities of each level of emergency responder.*

- *Describe six events that would trigger an emergency response.*

- *List four teams that assist in emergency response.*

- *Describe each of the emergency incident zones.*

- *Describe what activities take place in the emergency incident zones.*

- *Explain the benefit of a post-incident analysis.*

INTRODUCTION

The Occupational Safety and Healthy Administration (OSHA) issued a special regulation dealing with chemical spills. The standard, 29 CFR 1910.120, is called the **Hazardous Waste Operations and Emergency Response, or HAZWOPER**. HAZWOPER gives organizations two options for responding to a chemical spill. The first is to evacuate all employees in the event of a spill and call in professional emergency response personnel. Employers who use this option must have an *emergency action plan* (EAP) in place in accordance with 29 CFR 1010.38(a). The second option is to respond internally. Employers who use this option must have an emergency response plan in place that is in accordance with 29 CFR 1010.120. Both of these are explained below.

EAPs are plans that have at minimum the following elements: alarms systems, evacuation plan, a mechanism or procedure for emergency shutdown of the equipment, and a procedure for notifying emergency response personnel.

Emergency response plans detail how companies choose to respond to spills. If they choose to respond internally to chemical spills, they must have an emergency response plan that includes the provision of comprehensive training for employees. HAZWOPER specifies the type and amount of training required, ranging from awareness to in-depth technical training for employees who will actually deal with the spill. OSHA forbids the involvement of untrained employees in responding to a spill. If the company chooses to respond externally, they will contract a company to handle an emergency response. Why is emergency response training necessary? The answer is revealed in Table 18-1, which shows large releases from refineries are not infrequent. Bear in mind that many of the releases were caused by bad weather events.

HAZARDOUS WASTE OPERATIONS

The acronym HAZWOPER is often used when referring to OSHA's Hazardous Waste Operations and Emergency Response standard. The standard covers two important parts

Table 18-1 Toxic Chemical Releases by Industry

1. Barge ran agrounds in Natticoke River near Chrisfield, Maryland
2. Gasoline leaked from a facility in Los Angeles, California
3. Diesel, gasoline, and jet fuel leaking from a pipeline in Elmira, California
4. Crude oil spilled from a pipeline in Woodsboro, Texas
5. Diesel oil spilled from a pipeline in Port Arthur, Texas
6. A storage tank in Lisbon, Louisiana, released crude oil
7. Crude oil and brine water mixture spill in Elk City, Oklahoma
8. Crude oil spill from storage tank in Longview, Texas
9. Crude oil spill from storage tank in Paradis, Louisiana
10. Gasoline spilled from a pipeline in Billings, Montana

© Cengage Learning 2013

of a plant's operation: emergency response and hazardous waste operations. It requires that all individuals who respond to an emergency situation have at least 24 hours of training. Refresher training is also covered under the standard while plant specific requirements may include additional training. HAZWOPER can be divided into three broad categories:

- Emergency response roles (four levels)—a first responder awareness level, first responder operations level, hazardous materials technician, and specialist level.
- Hazardous waste operations—which includes an incident command system, scene safety and control, spill control and containment, decontamination procedures, and all clear.
- Hazard protection—which involves prevention and control-terms and definitions, personal protection equipment (PPE) levels, identifying hazardous materials, hazards initiating an emergency response, avoiding hazards, entry of hazardous materials into the body, unit monitors, and field survey instruments.

An incidental release is a release of a hazardous substance which does not pose a significant safety or health hazard to employees in the immediate vicinity or to the employee cleaning it up, nor does it have the potential to become an emergency within a short time frame. Incidental releases are limited in quantity, exposure potential, or toxicity and present minor safety or health hazards to employees in the immediate work area or those assigned to clean them up. An incidental spill may be safely cleaned up by employees who are familiar with the hazards of the chemicals with which they are working.

The chemical manufacturing industry defines **emergency response** as a loss of containment for a chemical or the potential for loss of containment that results in an emergency situation requiring an immediate response. Examples of emergency response situations include fires, explosions, vapor releases, and reportable quantity chemical spills. Refinery and petrochemical plant employees who are likely to discover or witness a chemical release fall under the scope of the awareness level while those employees who take preventive measures to control and secure the release fall under the guidelines of the operations level. In the refining and petrochemical industries, emergency response procedures are generally applied to every individual working for a company.

The HAZWOPER standard defines five levels for emergency responders. These five levels include:

1. First responder awareness level
2. First responder operations level
3. Hazardous materials technician level
4. Hazardous materials specialist
5. Incident commander

In this section of this chapter, we will look only at the roles of the first responder awareness level and first responder operations level. Further in this chapter, we will look at the role of the incident commander. Hazardous materials technician and specialist levels will not be covered. Readers wanting to know about those roles should consult 29 CFR 1910.120.

First Responder Awareness Level

Everyone working in the plant fits into this emergency response category and receives training for this level. The first responder awareness level is directed at individuals who witness or discover a hazardous chemical release. Properly trained personnel know how to recognize a hazardous chemical release, the hazards associated with this release, how to initiate the emergency response procedure, and how to notify appropriate personnel. A clerk working in the administration building but delivering records to a process unit might drive past a pipeline or tank that is leaking. Due to their first responder training, they would recognize that a release has occurred and notify appropriate personnel. The point being made is that anyone working within the plant can and should receive first responder awareness training.

First responder awareness training consists of:

- How to recognize a hazardous chemical release. Is a stream of liquid running across the road water or a hazardous chemical? Is the appearance of a white cloud steam or a hazardous vapor release?
- How to recognize the hazards associated with the release. Is the release coming from a tank, pipe, or pump? Did you detect an odor or color? Did you see identifying symbols or numerals on the leaking vessel? What is the location of the release?
- How to initiate the emergency response procedure (emergency phone number or radio frequency or alert area operators).
- How to notify appropriate personnel of an emergency response.

Personnel who receive emergency response awareness level are required to have refresher training annually. The training is documented.

First Responder Operations Level

Process technicians and engineers and working on the process units receive this training upon initial assignment and annually each year thereafter. When a hazardous chemical release occurs, the process technicians working on the unit specific to the release will respond to the emergency. They are trained to respond in a defensive manner without actually trying to stop the release. Their goals are to control or contain the release and protect lives, equipment, and the environment. First responders at the operations level learn:

- To control and contain the release
- The structure of emergency response system
- How to use the appropriate personal protective equipment
- Standard emergency response and termination procedures

If you think about it, who best knows where the shutoff valves on a unit are, what chemicals are in what lines and vessels, and where the unit's emergency equipment is located than the operators of the affected unit? This is why most, if not all, process technicians are designated First Responder, Operations Level.

EMERGENCY RESPONSE

All safety programs are designed to prevent or eliminate unsafe incidents. However, in view of the number of chemical and physical hazards that exist in refineries, petrochemical plants, and other types of processing industries it is recognized that unsafe events can occur. Therefore, there should be a strategy in place to minimize the consequences of explosions, fires, spills, and gas releases. Many regulations, including HAZWOPER, have emergency plans as an integral component.

Emergencies happen quickly. The function of an emergency plan (contingency plan) is to put in place resources necessary to deal with emergencies. Since there are all kinds of emergencies, the planning has to be sufficiently broad to cover any type of emergency. It is also important to note that some generalized response elements are common for many different emergencies. Some of the elements of an emergency plan are:

- Pre-emergency preparations
- Lines of authority
- Communication channels
- Notification procedures
- Emergency recognition
- Shelters
- Evacuation routes
- Site security
- Decontamination plans
- Emergency medical treatment
- PPE
- Post emergency evaluation

The emergency response plan is not a static document. It should be reviewed and modified periodically to reflect changing technology, changes in raw materials and products, and changes in regulations. The emergency response code signals should be easy to understand by all site personnel, and should be brief and loud enough so that it can be heard by all.

Emergency Response Situations

Some events that would trigger an emergency response (see Figure 18-1) are:

- The loss of containment of a hazardous substance
- A punctured 55-gallon drum
- A pump or compressor seal failure
- An overflowing tank
- A pipe or vessel leak
- An explosion or fire
- A gas release or vapor release
- A toxic chemical spill or release

The primary causes for loss of containment of fluids are pipe or flange failure, pump seal failure, explosions, fires, overfilled tanks, over-pressured tanks, and overturned drums or containers. Because loss of containment creates serious situations, the early

© Cengage Learning 2013

Figure 18-1   Emergency Response Railcar Leak

detection of loss of containment is important. Besides a loss of containment being detected by working personnel via sight and odor, fixed field monitors are installed around especially hazardous chemical vessels. Fixed unit monitors are designed to test for combustible or toxic substances. Toxic substances might be chemicals such as hydrogen sulfide or chlorine. Organic vapor analyzers and oxygen analyzers are used to detect hydrocarbons and oxygen levels. Fixed detectors are strategically located around units, tank farms, and dock areas. They send an audible alarm to the control room in the event a target substance is detected.

People Component of an Emergency Response Team

In preparing to manage emergencies, the organization of personnel and their roles should be defined. Emergency response personnel include both onsite as well as offsite individuals. The emergency response plan should define a chain of command. The plan should be well defined and yet flexible enough to respond to different types of emergencies, or even multiple emergencies.

The **Incident Commander (IC)** is the leader that directs the emergency response actions. Typically, in chemical plants and oil refineries, a shift supervisor acts as the incident commander, but this is not always the case. In emergencies, decisions have to be made quickly, thus the IC should have the full support of management. The IC must have a backup in case they are on vacation, in a hospital, etc. The IC should also have the authority to purchase supplies, to manage people and supplies entering/leaving the site, and to resolve disputes.

Personnel actually working to contain or control the release are the first responder operations level (previously discussed), the hazardous materials technician level, and the hazardous materials specialist.

Hazardous materials technicians are individuals who respond to releases or potential releases for the purpose of stopping the release. They assume a more aggressive role than the first responder. Many operators on processing units may have dual roles of first responder operations and hazardous materials technician. Hazardous materials technicians receive more training than first responder operations.

Hazardous materials specialists are individuals who respond with and provide support to hazardous materials technicians. Their duties parallel the technician's; however, their duties require a more specific knowledge of the chemicals they may be called upon to contain. They often act as the site liaison with local, state, or federal authorities in regard to the emergency.

To support the IC, there are several onsite teams given specific tasks. Examples are a *communication team, medical assist team, decontamination team, a team to report to regulatory agencies, a team to monitor onsite and offsite exposures, rescue team,* and other teams as necessary. Each team should have a point person who is responsible for coordinating the activities of the team.

Some plants rely on outside teams (contractors) to supply specialized functions such as rescue and shelter. Obviously, the time to arrange for outside contractors is before an emergency occurs. In selecting the offsite services, site mangers consider numerous factors, including the location of the offsite service (the company should be in close proximity of the plant), traffic on the highways, backup facilities, their insurance, and references from other plants in the area.

COMPONENTS OF AN EMERGENCY RESPONSE PLAN

The emergency response plan should be well thought out and reviewed constantly to ensure nothing has been left out or whether modifications to the plan are needed due to new chemicals or process units in the plant. Some of the more critical components (see Figure 18-2) are discussed in this section.

Communication

It is important that the public, neighbors, and news media are kept informed of an incident, if warranted. The public relations person can be given information by the IC or by his representative. Outside agencies or news organizations should contact the communication person and not just anybody in the plant. For the information from the plant to be as accurate as possible, employees should be instructed to direct inquiries from outside persons to the communication team. This guideline that employees should direct media phone calls to the communications group does not mean that a company seeks to hide information.

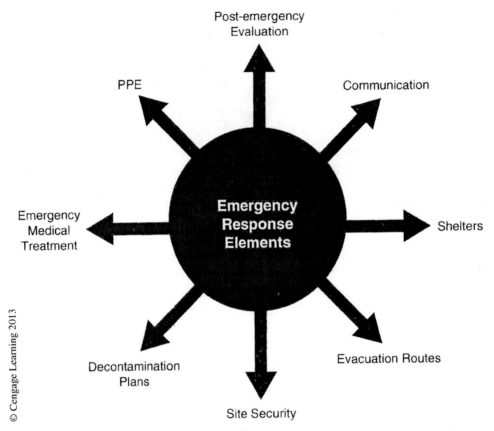

© Cengage Learning 2013

Figure 18-2 Emergency Response Plan Elements

Rather, it stresses the importance of the incident information being presented in a unified manner. The person communicating to the news media should be able to answer questions clearly without jumping to conclusions or making sensational or alarmist statements. The public perception created with just a few minutes of TV or radio interview could have a lasting effect on the worker's plant site—for good or bad. Many plants maintain a bank of dedicated computer networks and phones and faxes and a specially trained team of people that facilitate the flow of information to the public and news media.

Offsite Personnel Training

Offsite personnel that are part of the emergency response team must be familiar with plant:

- Hazards
- Road systems (plant plot plan)
- Safety procedures
- Site specific hazards
- Emergency response techniques

For in-plant personnel, and especially for offsite persons, the map or the plot plan is an effective way to describe the size and the location of the site. It also helps locate where the emergency has developed. When the risk management program (RMP) was being developed by the companies of the Gulf Coast area, computer-based maps were extensively used to display the areas in and around the plants.

Onsite Safety Stations

Onsite safety stations basically are small rooms which store some supplies for emergency responders and serve as meeting places for in-plant people to plan shift activities. They may also contain portable gas detectors, communication gear, first aid supplies, and PPE, especially self-contained breathing apparatus.

Site Security

Amid the commotion of the emergency site, security should not be forgotten. The site security team keeps track of people entering and leaving the plant. A signature roster or proximity card is a good way to keep track of workers onsite and visitors in the plant. The IC should have knowledge of the people entering the plant. Typical checks applied during normal operation can be extended during an emergency. Periodic head counts are carried out to account for everyone onsite.

Evacuation Routes

Some emergencies, like a fire, explosion, or even toxic cloud, may cut off people from their assembly areas. Therefore, routes of evacuation and alternate routes should be clearly understood by all site personnel and visitors. All persons entering the site should be given a thorough explanation of the routes and alternate means of exit. The alternate and primary route should lead in different directions from each other.

Emergency Equipment

The type of equipment required for an emergency varies from one emergency to another. For instance, some emergencies may require heavy equipment such as cranes, bulldozers, cherry pickers, etc. Although, most of the plants have fork lift trucks, man-lifts, and other medium sized-equipment, not all plants keep cranes or other heavy equipment onsite. Such specialized equipment may be necessary depending on the type of the emergency. A part of the emergency procedure should require a list of contractors, their phone numbers, locations, and availability of emergency equipment.

Medical Services and Equipment

Quick availability of first aid and cardiopulmonary resuscitation (CPR) can be critical. It is important to have a group of people on each shift that can administer CPR and render other medical assistance. Medical emergency supplies should be available at appropriate stations. Liaison should be established with the nearby medical facilities for online assistance.

Decontamination Plans

Persons coming out of the exclusion zone (discussed later in this chapter) must be decontaminated of hazardous chemicals on their gear before they enter any other zone. Decontamination has three basic objectives which are:

- Decontaminate the worker or victim.
- Provide protection of the medical personnel.
- Dispose of contaminated PPE and washings from the emergency event.

Whether to perform decontamination on a victim depends on his/her physical condition. If a person is seriously hurt, it may not be possible to decontaminate them at that time. In that case, wrap the victim in blankets and alert medical professionals that decontamination could not be performed. Facilities should be provided containers (55-gallon DOT approved drums may be suitable for many applications) to store the contaminated PPE and wash solutions which can be disposed of later.

EMERGENCY INCIDENT ZONES

Effective emergency response requires a good site control program. The purpose of a site control program is to minimize worker and public exposure to any hazards resulting due to an emergency. These hazards may be chemical, biological, or other safety related hazards. The level of site control depends on the size of the site, topography, and the neighboring community. Site preparation is a necessary first step to ensure effective emergency response. Roadways within the plant should be constructed to facilitate traffic to all areas of the plant.

To facilitate workflow, the areas around an emergency incident are divided into three zones: *exclusion zone, contamination reduction zone (CRZ)*, and *support zone* (see Figure 18-3). The zones should be located upwind of the emergency site. To minimize spread of contamination, workers should not cross from one zone to another except at approved points, called *access control points*.

Exclusion Zone

The area of the spill or actual emergency is called the **exclusion zone**. Formerly, it was referred to as the *hot zone*. This is the contaminated area and no unauthorized persons should enter this area. Site clean-up, drum movement, drum staging, and clean up are some of the activities conducted at the exclusion zone. Since this is a hazardous zone, the buddy system is mandatory in this zone. Workers should be constantly monitored visually. To ensure safety, all the workers in the exclusion zone should always be in the line of sight. The boundary of the exclusion zone is called the *hotline*. The hotline should be

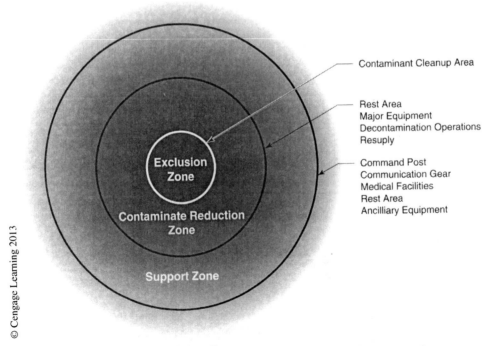

Contaminant Cleanup Area

Rest Area
Major Equipment
Decontamination Operations
Resuply

Command Post
Communication Gear
Medical Facilities
Rest Area
Ancilliary Equipment

Exclusion Zone

Contaminate Reduction Zone

Support Zone

© Cengage Learning 2013

Figure 18-3   Emergency Response Zones

clearly marked to show that unauthorized personnel are not allowed. SCBA may be sufficient for many situations; however, it is possible to have only cartridge masks if atmospheric hazards are known with certainty.

Contaminant Reduction Zone (CRZ)

The contamination reduction zone (CRZ) surrounds the exclusion zone. In a sense, this is a buffer zone between the exclusion zone and support zone. In the CRZ, there is a narrow path provided where decontamination (decon) is carried out and the path is called the *decon line*. By restricting the decon path, we localize the spread of contamination. All decontamination operations are carried out near or along the path, and closer to the end of the path, leading to the support zone. The CRZ is a rest area for workers and also serves as the staging area for the emergency response equipment. Typically, the CRZ is where decontamination of personnel and equipment is carried out. Some activities that occur in the CRZ are:

- Major equipment is stationed there (PPE, air supply bottles, SCBA, etc.).
- Sample handling (for laboratory analysis) occurs here.
- It is a temporary rest area for workers.
- First aid treatment occurs here.
- Drainage and containment of waste fluids resulting from the decon operation are captured and stored here.

Although the CRZ is less hazardous than the exclusion zone, just like in the exclusion zone, people in the CRZ should be in line of sight and must maintain constant communication. The decontamination facility should be so situated as to minimize the spread of contaminants. The facilities should be designed for quick response to decon requirements. The decon facility design should take the following into consideration:

- Properties of the wastes (toxic, flammable, chemical, etc.)
- Amount, location, containment, and storage of contaminants
- Potential for exposure
- Located away from a public waterway
- The movement of workers

Decon procedures (see Figure 18-4) should be designed to limit worker exposure and gradually reduce the level of contaminants on the PPE of the workers. It may consist of two walls of sprays (showers) in series, doffing off the PPE and provision of new PPE. The decon washings are stored in DOT approved containers. Depending on the nature of the emergency, there may be several decon stations. The stations closest to the exclusion zone would require the workers to wear the same level of protection as the exclusion zone workers themselves. As workers move away from the exclusion zone, the level of protection may gradually decrease from A to B down to C. The six basic decontamination steps may be divided between six decontamination stations.

- Station 1: Equipment drop off (respirators, tools, etc.)
- Station 2: Wash and rinse boots and gloves
- Station 3: SCBA tank change
- Station 4 Boots, gloves, and outer garment removal
- Station 5: SCBA facepiece removal
- Station 6: Field wash

239

© Cengage Learning 2013

Figure 18-4 Decontamination of Responder

Support Zone

The support zone is the last zone surrounding the exclusion zone. This zone contains people not directly involved in eliminating or controlling the emergency. This zone contains administrative personnel, a communication station, food and drinks, and other accessories. This is the area provided with all the vital information about the chemicals on site, site topography, and other resources as needed. The incident commander's command post is also located in the support zone. This is also a staging area for equipment and PPE to be moved to the CRZ. The following is a typical list of the personnel and equipment found in the support zone:

- Administrative personnel
- Command post
- Engineering files and procedural manuals
- Medical facilities
- Communication gear
- Laboratory for field analyses
- Worker rest areas: snacks, drinks, wash facilities, toilets, etc.
- Ancillary equipment (cherry picker, forklifts, crane on standby, etc.)

Demarcation of zones depends on many factors including the size and type of emergency, wind direction, and terrain. The support zone location and orientation is governed by:

- Easy access, close to highways, and railroads
- Availability of utilities

- Visibility—from the command center, all the teams involved in the emergency work should be visible
- Wind direction—the support zone (in fact all zones) should be upwind from the site of emergency

Communication and Documentation

While the emergency is in progress, several outside agencies and local government departments will be notified of the nature of the emergency. This might involve the notification of federal and/or state regulatory agencies with jurisdiction over environmental matters. Notification to local agencies (city hall, fire, and police departments), and a neighbor notification network (CAER). For instance, Houston and the Gulf Coast area have CAER, a phone-based network that can automatically dial multiple locations simultaneously and give a brief message of the emergency.

Post-Incident Analysis

Accurate and timely documentation of all emergency responses is vital for two reasons. First, documentation provides valuable information for future courses of action. It reveals what went right and what went wrong. The second reason is for legal considerations. Documented data could be a valuable resource should a lawsuit arise later. We can always improve upon anything we do. This is the basic guiding principle of the post-incident analysis. Maybe a responder had to wait 30 minutes for a fresh SCBA tank. Or the IC could not answer a TV reporter's question about a ruptured storage tank's last preventive maintenance. At the end of every emergency, a meeting should be held to discuss or critique the incident. The purpose of incident analysis is to improve emergency response procedures and planning, not to find fault. This session should be free of company politics and finger pointing.

SUMMARY

OSHA issued a special regulation dealing with chemical spills. The standard, 29 CFR 1910.120, is called the Hazardous Waste Operations and Emergency Response, or HAZWOPER. The standard covers two important parts of a plant's operation: emergency response and hazardous waste operations. Emergency response roles consist of five levels—first responder awareness level, first responder operations level, hazardous materials technician, specialist level, and incident commander. Hazardous waste operations consists of the incident command system, scene safety and control, spill control and containment, decontamination procedures, and the all clear.

The five levels of emergency responders all have varying degrees of training, from minimal to very intense. When a spill occurs that requires an emergency response, the area around the spill is quickly divided into three zones to facilitate workflow. These zones are the exclusion zone, contamination reduction zone and support zone. In each zone, specific actions are carried out. An effective emergency response is based on quickness of response, properly trained personnel, proper equipment, and how efficiently they are used.

REVIEW QUESTIONS

1. Describe the two options HAZWOPER gives organizations for responding to a chemical spill.

2. Two important parts of a plant's operations that HAZWOPER covers are:
   a. Community awareness of hazards
   b. Emergency response
   c. Firefighting systems
   d. Hazardous waste operations

3. (T/F) OSHA forbids the involvement of untrained employees responding to a spill.

4. An_____ release is a release of a hazardous substance which does not pose a significant safety or health hazard to employees in the immediate vicinity.

5. List the five levels of emergency responders.

6. (T/F) Everyone working in a plant fits into the first responder awareness level.

7. (Pick three) Properly trained first responders know how to:
   a. Recognize a hazardous chemical release
   b. Block in valves to shutoff the spill
   c. Recognize hazards associated with the release
   d. Alert the plant manager
   e. Notify appropriate personnel

8. People who receive emergency response awareness level training are required to have refresher training _____.

9. _____ is a loss of containment for a chemical or the potential for loss of containment that results in an emergency situation.

10. List six events that would trigger an emergency response.

11. Describe the duties of the first responder operations level.

12. List four teams that assist in emergency response.

13. The _____ is the leader who directs the emergency response actions.

14. (Circle those that apply) Communication to local agencies and news media during an emergency response is important because:
   a. Public perception can be affected
   b. It prevents jumping to conclusions
   c. It is required by OSHA

15. _____ are small rooms which store some supplies for emergency responders.

16. Persons coming out of the exclusion zone must be decontaminated of hazardous chemicals on their gear before entering another zone. Decontamination has **three** objectives:
   a. Decontaminate the worker or victim

   b. Prevent fugitive emissions from the gear
   c. Provide protection of medical personnel
   d. Dispose of contaminate PPE and washings

17. The areas around an emergency incident are divided into three zones, which are the
    _____, _____, and _____.

18. The _____ is a rest area for workers and also serves as the staging area for the
    emergency response equipment.

19. The _____ is the outer zone of the emergency site and contains people not
    directly involved in eliminating or controlling the emergency.

20. (Circle all that apply) The benefits of a post-incident analysis are:
    a. Valuable information for future courses of action
    b. Legal considerations
    c. Community right to know
    d. Required by OSHA

EXERCISES

1. **Instructions:** Choose one of the following environmental disasters where hazardous
   substances were released into the environment. Research the accident and write a
   two page report that (1) explains the cause of the accident, (2) the short term effects
   of the accident, and (3) the long term effects of the accident.
   a) Love Canal
   b) Chernobyl (Russia)

2. Research the Internet for a PowerPoint presentation on HAZWOPER and copy it to
   a flash drive or CD. Bring it to class to be presented to the class by yourself or the
   instructor. Cite your source.

RESOURCES

www.wikipedia.org

www.osha.gov

www.YouTube.com

tceq.state.tx.us

www.toolboxtopics.com

CHAPTER 19

The Occupational Safety and Health Administration (OSHA) and U.S. Department of Transportation (DOT)

Learning Objectives

Upon completion of this chapter the student should be able to:

- *Explain the reason for OSHA being created as an agency.*

- *Explain OSHA's reporting requirements.*

- *Describe three things employers are required to do in order to keep employees informed.*

- *List and explain three different types of OSHA citations and the typical penalties that accompany them.*

- *Describe five employer responsibilities under OSHA.*

- *Describe five employee rights under OSHA.*

- *List five employee responsibilities under OSHA.*

- *State the primary function of the U.S. Department of Transportation (DOT).*

INTRODUCTION

The Occupational Safety and Healthy Administration (OSHA) was created in 1970 with a mission of attempting to reduce the number and severity of accidents by making equipment and procedures safer by mandatory means. Since 1971, workplace fatalities have steadily fallen, an accomplishment that might not have been achieved without OSHA. Workplace injuries continue to decline despite the increasing number of hours worked.

The application of only OSHA regulations in the workplace rarely results in an accident and injury status of zero. OSHA and its regulations (or fear of OSHA) are not enough to have a successful safety program. OSHA regulations are essential but must be combined with other safety practices, such as:

- Planned safety inspections
- Job hazard analyses
- Incident investigations
- Job safety observations

Besides the above, machines have to be guarded, housekeeping kept intact, chemicals safely used and stored, personal protective equipment properly worn, and correct lighting and signage used. Employee involvement is also essential.

Until a few years ago, it was widely believed that if a worker was involved in an accident while performing their duties, it was probably the worker's fault. In a review of 75,000 accident cases, an estimated 88 percent of the accidents were due to unsafe acts of people. Similarly, airplane accidents were generally assumed due to human error—the pilot. U.S. Air Force ballistic missiles introduced a new concept of the cause of accidents. When the missiles were launched and then failed, there was no human on board to blame. Thus, the cause of the accident was due to something else, such as design or manufacturing error. This led to the new concept that accidents could be caused by an error due to:

- Equipment design
- Management
- Maintenance
- Environment
- Anyone connected with the process or system

In the previous paragraph, it was mentioned that in the past 88 percent of accidents were due to unsafe acts by the worker. Today, if those accidents were re-examined using the new concept, the data might say that 88 percent of the accidents were due to causes other than human error. This concept led to the OSHA Act, which attempts to reduce the number and severity of accidents by making equipment and procedures safer by mandatory means.

The three primary agencies responsible for the administration of the OSHA Act (see Figure 19-1) are briefly described in the following bulleted sentences.

- OSHA is charged with investigating catastrophes and fatalities in the workplace, inspecting workplaces, and establishing new safety and health standards and penalties.
- National Institute for Occupational Safety and Health (NIOSH) is charged with developing toxic substance handling criteria, researching safety and health issues, recommending new safety, and health standards.
- Occupational Safety and Health Review Commission (OSHRC) is charged with conducting hearings for situations on noncompliance that are contested, and supporting, modifying, or overturning OSHA findings.

Figure 19-1   Three Agencies of OSHA

OSHA IS BORN

Despite the progress made in identifying and controlling occupational illnesses and injuries, many workers still face serious health and safety hazards in the workplace. Though the number of occupation deaths has declined in recent years, the number of work-related injuries and illnesses appear to be increasing. It is only recently that occupational disease, particularly those of long latency, has been recognized as being occupationally related. It takes years or decades to see a positive correlation pattern develop between long latent occupational diseases and the worker's occupation.

Growing pressure from unions and industries eventually led to the passage of the Williams-Steiger Occupational Safety and Health Act in 1970 with the objective to "assure so far as possible every man and woman in the nation safe and healthful working conditions." When congress developed the Occupational Safety and Health Act, it did so after consideration of the following statistics:

- Each year, an average of 14,000 workplace deaths occurred.
- Each year, 2.5 million workers were disabled in workplace accidents.
- Each year, approximately 300,000 new cases of occupational diseases were reported.

A comprehensive set of regulations was needed to help reduce the incidence of work-related injuries, illnesses, and deaths. The OSHA Act of 1970 addressed this need with Title 29 of the Code of Federal Regulations, Parts 1900 through 1910. The Act also established the OSHA, which is part of the U.S. Department of Labor, and is responsible for administering the OSHA Act.

OSHA's mission and purpose is very comprehensive, as revealed by the following list.

1. Encourage employers and employees to reduce workplace hazards.
2. Implement new safety and health programs.
3. Improve existing safety and health programs.
4. Encourage research that will lead to innovative ways of dealing with workplace safety and health problems.
5. Establish the rights of employers regarding the improvement of workplace safety and health.
6. Establish the rights of employees regarding the improvement of workplace safety and health.
7. Monitor job-related illnesses and injuries through a system of reporting and record keeping.
8. Establish training programs to increase the number of safety and health professionals and to improve their competence continually.
9. Establish mandatory workplace safety and health standards and enforce those standards.
10. Provide for the development and approval of state-level workplace safety and health programs.
11. Monitor, analyze, and evaluate state-level safety and health programs.
12. Encourage joint labor-management efforts to reduce injuries and disease arising out of employment.

Application of the OSHA Standard

The OSHA Act prescribes minimum safety and health standards for most industrial sectors and authorizes inspections by OSHA inspectors. For enforcement purposes, the OSHA Act allows citations (notice of violation) and civil penalties. It also allows states to have more stringent standards for their own OSHA-type agencies and to administer their own occupational, safety, and health program.

Section 8(c) of the OSHA Act requires employers to maintain a record of work-related illnesses and injuries in an OSHA 300 Log. Upon request, this record must be made available to employees, former employees or their representatives. Section 2 (b) also places responsibility on workers by encouraging them to (1) reduce the number of occupational safety and health hazards in their workplace, and (2) to strive to achieve safe and healthful working conditions.

OSHA does not inspect businesses with ten or fewer employees unless (1) a fatality has occurred, or (2) there has been an employee complaint. The OSHA Act applies to most employers involved in manufacturing, construction, retail, and service organizations even if they have just one employee because there are applicable sections of the Act they must comply with. Although there is no exemption for small businesses from the OSHA Act, organizations with ten or fewer employees are exempted from OSHA inspections and the requirement to maintain injury/illness records.

There are more than 97 million workers in the United States. Businesses with fewer than ten employees make up 74 percent of all businesses and 16 percent of the workforce.

These figures reveal that a significant portion of workplaces are not subject to OSHA inspections. Also, businesses with 11-50 employees make up 16 percent of businesses and 24 percent of the workforce. Due to the limited number of OSHA inspectors, businesses of this size are rarely inspected. Thus, OSHA does not regularly inspect approximately 90 percent of work sites.

In general, the OSHA Act covers employers in all 50 states, the District of Columbia, Puerto Rico, and all other territories that fall under the jurisdiction of the U.S. government. Exempted employers are:

- Self-employed
- Family farms employing only immediate members of the family
- Federal agencies covered by other federal statutes (in cases where these other federal statutes do not cover working conditions in a specific area or areas, OSHA standards apply)
- State and local governments
- Coal mines, which are regulated by mining-specific laws

OSHA Standards

In carrying out its duties, OSHA is responsible for declaring legally enforceable standards (laws). OSHA standards may require (1) the changing of working conditions, and/or (2) the adoption of new practices, methods, or processes reasonably necessary to protect workers on the job.

The *general duty clause* of the OSHA Act requires that employers provide a workplace free from hazards likely to harm employees. This clause is important because it applies when there is no specific OSHA standard for a given situation. This clause can be used when a hostile environment exists, such a worker threatening another worker. Where OSHA standards do exist, employers are required to comply with them as written. It is the responsibility of employers to become familiar with standards applicable to their businesses. Table 19-1 lists the ten most frequently cited standards.

How OSHA Standards Are Developed

OSHA develops standards based on its perception of need and at the request of other federal agencies, state and local governments, and labor organizations. OSHA uses committees to develop standards. Committees can be standing committees within OSHA and special ad hoc committees appointed to deal with issues that are beyond the scope of the standing committees. NIOSH, which is a separate agency under the OSHA Act, has an education and research orientation. The results of this agency's research are often used to assist OSHA in developing standards.

OSHA can adopt, amend, or revise standards but before any of these actions can be undertaken, OSHA must publish its intentions in the *Federal Register* in either a notice of proposed rule making or an advance notice of proposed rule making. The notice of proposed rule making must explain the terms of the new rule, point out proposed changes to existing rules, or list rules that are to be revoked. After publishing the notice, OSHA must conduct

Table 19-1 10 Most Frequently Cited Standards

Subject	Standard
1. Scaffolding	1926.451
2. Fall Protecting	1926.501
3. Hazard communication—labeling containers	1910.1200
4. Respiratory Protection	1910.134
5. Ladders	1926.1052
6. Lockout/Tagout	1910.147
7. Electrical, Wiring Methods	1910.305
8. Powered Industrial Trucks	1910.178
9. Electrical, General Requirements	1910.303
10. Machine Guards	1910.212

© Cengage Learning 2013

Source: Federal Data, as of October 8, 2010, www.osha.gov

a public hearing if one is requested. Any interested party may ask for a public hearing on a proposed rule or rule change. OSHA must schedule the hearing and announce the time and place in the *Federal Register.* The final step occurs after the close of the comment period and public hearing. OSHA must publish in the *Federal Register* the final text of any standard amended or adopted and the date it becomes effective, along with an explanation of the standard and the reasons for implementing it. OSHA may also publish a determination that no standard or amendment needs to be issued.

WORKPLACE INSPECTIONS

When OSHA inspectors go into a workplace for the purpose of an inspection, they are trying to determine if the company:

- Keeps employees informed of rules and regulations
- Enforce its rules and regulations
- Provides its employees with the necessary training

OSHA uses workplace inspections for enforcing its rules. OSHA personnel may conduct workplace inspections unannounced. Except under special circumstances, it is a crime punishable by fine, imprisonment, or both to give an employer prior notice of an inspection. When OSHA compliance officers arrive to conduct an inspection, they are required to present their credentials. This authorizes them to enter any site or facility where work is taking place. They may inspect, at reasonable times, any condition, facility, machine, equipment, materials, etc. They may question in private any employee or other person formally associated with the company. Under special circumstances, employers may be given up to a maximum of 24-hours notice of an inspection.

OSHA has limited resources, and because the OSHA Act applies to approximately six million work sites in the United States, OSHA must establish priorities for which sites get inspected. These priorities are:

- Sites with imminent danger situations
- Sites with catastrophic fatal accidents
- Sites with employee complaints
- Planned high-hazard inspections
- Follow-up inspections

After an inspection is scheduled, the OSHA compliance officer presents their credentials to a company official and conducts an opening conference with pertinent company officials and employee representatives. During the conference, the following information is explained:

- Why the plant was selected for inspection
- The purpose of the inspection
- The inspection's scope
- The applicable standards

By proactively targeting the industries and employers that experience the greatest number of workplace injuries and illnesses, OSHA continues to maintain its high level of annual inspection activity. In fiscal year 2009, OSHA conducted 39,004 total inspections. OSHA conducted 24,316 programmed inspections and 14,688 unprogrammed inspections, including employee complaints, accidents, and referrals. The number of fatality investigations decreased by 28.5 percent; a significant decrease over the past five fiscal years (see Table 19-2).

Table 19-2   OSHA Inspection Statistics

	FY2005	FY2006	FY2007	FY2008	FY2009	%Change 2005–2009
Total Inspection	38,714	38,579	39,324	38,667	39,004	75%
Total Programmed Inspections	21,404	21,506	23,035	23,041	24,316	13.6%
Total Unprogrammed Inspections	17,310	17,073	16,289	15,626	14,688	−15.1%
Fatality Investigations	1,114	1,081	1,043	936	797	−28.5%
Complaints	7,716	7,376	7,055	6,708	6,661	−13.7%

© Cengage Learning 2013

Source: Material, data, and tables are provided by the U.S. Bureau of Labor Statistics, U.S. Department of Labor, and Occupational Safety and Healthy Administrations' area offices.

The compliance officer makes the inspection tour, and during the tour, may observe, interview pertinent personnel, examine records, take readings, and make photographs. Later, the compliance officer holds a closing conference, which involves open discussion between the officer and company/employee representatives. OSHA personnel advise company representatives of problems noted, actions planned as a result, and assistance available from OSHA.

CITATIONS AND PENALTIES

An OSHA citation informs the employer of OSHA violations. Penalties are fines assessed as the result of citations. In March 1991, civil monetary penalties were increased sevenfold. The maximum allowable penalty is $70,000 for each willful or repeated violation, and $7,000 for each serious or other-than serious violation as well as $7,000 per day beyond a stated abatement date for failure to correct a violation. Examples of very large penalties paid out are:

- USX Corporation agreed to pay a penalty of $3.25 million.
- ARCO settled the citation and paid $3.48 million that followed a disastrous explosion in Texas resulting in the deaths of 17 employees.
- BASF Corporation agreed to pay the $1.06 million penalty after an explosion in Cincinnati resulted in the death of two employees.

With OSHA penalties increasing in dollar amounts, there is more incentive for employers to appeal or even contest them through litigation. See Table 19-1 for a list of the most frequent violations of OSHA standards.

Some of the types of citations and their corresponding penalties are described below.

1. **Other-than-serious violations** are violations that have a direct relationship to job safety and health, but probably would not cause death or serious physical harm. A proposed penalty of up to $7,000 for each violation is discretionary.
2. **Willful violations** are violations that the employer intentionally and knowingly commits. The employer either knows that what they are doing is a violation or is aware that a hazardous condition exists and has made no reasonable effort to eliminate it. Penalties of up to $70,000 may be proposed for each willful violation, with a minimum penalty of $5,000 for each violation.
3. **Repeat violations** are violations of any standard, regulation, rule, or order where, upon reinspection, a substantially similar violation is found. Repeat violations can bring a fine of up to $70,000 for each such violation.
4. **Failure to correct prior violations** may bring a civil penalty of up to $7,000 for each day that the violation continues beyond the specified abatement date.

Employers may also be penalized by additional fines and/or prison if convicted of any of the following offenses:

1. Falsifying records or any other information given to OSHA personnel.
2. Failing to comply with posting requirements.
3. Interfering in any way with OSHA compliance officers in the performance of their duties.

Section 17(e) of the OSHA Act provides criminal penalties for an employer who is convicted of willfully violating an OSHA standard, rule, or order when the violation results in the death of an employee. Cases that may be appropriate for criminal prosecution are referred to the U.S. Department of Justice. In 2006 there were 12 criminal referrals, 2007 had 10, 2008 had 14, and 2009 had 11 (data from www.osha.gov, October 2010).

THE APPEALS PROCESS

Employee Appeals — Workplace employees may not contest the fact that citations were or were not awarded or the amounts of the penalties assessed their employer. However, they may appeal the following aspects of OSHA's decisions regarding their workplace and the appeals must be filed within ten working days of a posting:

1. The amount of time (abatement period) given an employer to correct a hazardous condition that has been cited.
2. An employer's request for an extension of an abatement period.

Employer Appeals — Employers may appeal a citation, an abatement period, or the amount of a proposed penalty. Formal appeals are of two types: (1) a petition for modification of abatement, or (2) a notice of contest.

A petition for modification of abatement (PMA) is available to employers who intend to correct the situation for which a citation was issued, but who need more time. As a first step, the employer must make a good-faith effort to correct the problem within the prescribed timeframe. Having done so, the employer may file a petition for modification of abatement.

An employer who does not wish to comply may contest a citation, an abatement period, and/or a penalty by notifying OSHA's area director in writing within 15 working days of receipt of a citation or penalty notice. The notice of contest must clearly describe the basis for the employer's challenge and contain all of the information about what is being challenged. Once OSHA receives a notice of contest, it is forwarded to the Occupational Safety and Health Review Commission (OSHRC). The commission assigns the case to an administrative law judge.

RESPONSIBILITIES UNDER OSHA

Both the employer and the employee have responsibilities under OSHA, and both are subject to civil and criminal penalties.

Employer Responsibilities

Employers have several responsibilities under the OSHA Act (see Figure 19-2), many of which, if they are in noncompliance, can lead to citations and penalties. What follows is a partial list of employer responsibilities.

1. Meet the general duty responsibility to provide a workplace free from hazards that are causing or are likely to cause death or serious physical harm to employees, and comply with standards, rules, and regulations issued under the OSHA Act.
2. Be knowledgeable of mandatory standards and make copies available to employees for review upon request.

Figure 19-2 Some Employer Responsibilities Under OSHA

3. Keep employees informed about OSHA.
4. Minimize or reduce hazards.
5. Ensure employees have and use safe tools and equipment that is properly maintained.
6. Use color codes, posters, labels, or signs as appropriate to warn employees of potential hazards.
7. Establish or update operating procedures and communicate them so that employees follow safety and health requirements.
8. Provide medical examinations as required by OSHA standards.
9. Provide training required by OSHA standards.
10. Report to the nearest OSHA office within 48 hours any fatal accident or one that results in the hospitalization of five or more employees.
11. Keep OSHA-required records of injuries and illnesses and post a copy of the totals from the last page of OSHA Form 300 during the entire month of February each year.
12. At a prominent location, post OSHA Poster 2203 informing employees of their rights and responsibilities.
13. Give employees access to medical and exposure records.

Employee Responsibilities

Under the OSHA Act, employees have the following specific responsibilities:

1. Read the OSHA poster at the job site and be familiar with its contents.
2. Comply with all applicable OSHA standards.
3. Follow safety and health rules and regulations prescribed by the employer and promptly use PPE while engaged in work.
4. Report hazardous conditions to a supervisor.
5. Report any job-related injury or illness to the employer and seek treatment promptly.
6. Cooperate with the OSHA compliance officer conducting an inspection.

MORE ABOUT OSHA

OSHA has a heavy burden investigating catastrophes and fatalities, establishing standards and penalties, and inspecting workplaces. That agency shifts some of its burden on the employer and employee.

Record Keeping

OSHA's mandatory record keeping requirements has simplified the process of collecting safety and health statistics for the purpose of monitoring problems and taking appropriate actions. Both exempt and nonexempt employers must report the following types of accidents within 48 hours: (1) those that result in deaths, and (2) those that result in the hospitalization of five or more employees.

All occupational illnesses and injuries must be reported if they result in one or more of the following:

- Death of one or more workers
- One or more days away from work for the employee
- Restricted motion or restrictions to the work an employee can do
- Loss of consciousness to one or more workers
- Transfer of an employee to another job
- Medical treatment needed beyond in-house first aid

Employee Rights

Employers cannot punishment an employee who takes any of the following courses of action:

1. Complain to an employer, union, OSHA, or any other government agency about job safety and health hazards.
2. File safety or health grievances.
3. Participate in a workplace safety and health committee or in union activities concerning job safety and health.
4. Participate in OSHA inspections, conferences, hearings, or other OSHA-related activities.

Employees who feel they are being treated unfairly because of actions they have taken in the interest of safety and health, have 30 days to contact the nearest OSHA office. Upon receipt of a complaint, OSHA will conduct an investigation and make recommendations based on its findings.

In addition to the rights just described, employees have the following rights:

1. Expect employers to make review copies available of OSHA standards and requirements.
2. Ask employers for information about hazards that may be present in the workplace.
3. Ask employers for information on emergency procedures.
4. Receive safety and health training.
5. Be kept informed about safety and health issues.

6. Anonymously ask OSHA to conduct an investigation of hazardous conditions at the work site.
7. Observe during an OSHA inspection and respond to the questions asked by a compliance officer.

Problems with OSHA

Because resentment of this federal bureaucracy is built into the American business man's mindset, complaints about OSHA are common. Some critics characterize OSHA as an overbearing bureaucracy with little sensitivity to the needs of employers who are struggling to survive in a competitive marketplace. Others label OSHA as timid and claim it does not do enough. Most criticism of OSHA comes in the aftermath of major accidents or a workplace disaster. The question often asked is, "Why didn't OSHA prevent this disaster?" OSHA detractors will claim that OSHA spends too much time and too many resources dealing with matters of little consequence while ignoring real problems. Supporters of OSHA will reply by claiming that a lack of resources prevents the agency from being everywhere at once. There is some truth in both answers.

An example of a disaster that precipitated this type of dialogue is the chemical explosion that occurred in May 1991 at a fertilizer plant in Sterlington, Louisiana. This incident claimed the life of eight employees and injured 128. OSHA had inspected the plant six times and issued eight different citations, but there had been no follow-up inspection of the plant in almost ten years. Charges of poor follow-up were leveled at OSHA. OSHA officials responded that they had followed up to the extent that their resources allow. The General Accounting Office (GAO), the investigative arm of Congress, found that OSHA often has to depend on the companies it cites to volunteer information about whether they have corrected the violations.

The reader should keep in mind that large, centralized bureaucratic agencies are rarely efficient. Also, OSHA is subject to the ebb and flow of congressional funding. Consequently, OSHA is likely to continue to be an imperfect organization.

Common Definitions

To prevent confusion and provide clarity, OSHA has quite a few definitions in its standards. These are few of the most common.

Negligence—means failure to take reasonable care or failure to perform duties in ways that prevent harm to humans or damage to property. The concept of *gross negligence* means (1) failure to exercise even slight care or (2) intentional failure to perform duties properly, regardless of the potential consequences.

Contributory Negligence—means that an injured party contributed in some way to his or her own injury. In the past, this concept was used to protect defendants against negligence charges because the courts awarded no damages to plaintiffs who had contributed in any way to their own injury. Modern court cases have rendered this approach outdated with the introduction of *comparative negligence*. This concept distributes the negligence assigned to each party involved in litigation according to the findings of the court.

Care—Several related concepts fall under the heading of *care*. *Reasonable care* is the amount that would be taken by a prudent person in exercising his or her legal obligations toward others. *Great care* means the amount of care that would be taken by an extraordinarily prudent person in exercising his or her legal obligations toward others. *Slight care* represents the other extreme: a measure of care less than what a prudent person would take. A final concept in this category is the *exercise of due care.* This means that all people have a legal obligation to exercise the amount of care necessary to avoid, to the extent possible, bringing harm to others or damage to their property.

Willful/Reckless Conduct—Behavior that is even worse than gross negligence is willful/reckless conduct. It involves intentionally neglecting one's responsibility to exercise reasonable care.

Foreseeability—The concept of foreseeability holds that a person can be held liable for actions that result in damages or injury only when risks could have been reasonably foreseen.

U.S. DEPARTMENT OF TRANSPORTATION (DOT)

The U.S. Department of Transportation (DOT) was established by an act of Congress on October 15, 1966. Huge quantities of flammable, combustible, and toxic materials are safely transported throughout the United States everyday. Though few accidents occur, when they do occur, they are more likely to affect the general public. One of the principal dangers of transportation accidents is the boiling-liquid-expanding-vapor explosion (BLEVE) that can happen if a trailer cargo tank or railcar is filled with a flammable liquid. If there is an accident and a fire occurs and contacts an unwetted portion of the shell of a railcar or cargo tank of a truck, the metal may fail catastrophically. When the shell fails, thousands of pounds of hot, flashing hydrocarbons are dumped into the fire and a gigantic fireball occurs. A BLEVE of a 30,000-gallon liquefied propane gas rail car will produce a fireball about 1,000 feet high and hundreds of feet in diameter.

In a nation that relies on transportation as heavily as the United States, the DOT assumes a variety of responsibilities. Principally, it strives to ensure transportation safety and protect consumer interests. One way it fulfills this mission is to require that cargo vessels are properly placarded and shipping papers properly filled out. Hazardous materials markings, labels, placards, and shipping papers serve to communicate the hazards of the materials being transported. Hazardous material communication:

- Is the key to effective emergency response.
- Alerts transportation workers and the general public of the presence of hazardous materials.
- Ensures that non-compatible materials are not loaded together in the same transport vehicle.
- Provides the necessary information for reporting hazardous materials incidents.

Process Operator Duties Under DOT

Hazardous materials are transported through and around business areas and neighborhoods. Tank trucks haul corrosive materials such as sulfuric acid and sodium hydroxide, and flammables like gasoline and liquefied petroleum gases. The cargo trailers hold approximately 30,000-gallons of material. Railroads chug through cities towing 50,000-gallon railcars of various flammable and toxic chemicals. A rupture of these cargo vessels can result in deaths, injuries, and the large-scale evacuation of business and residential areas. We have all been on a freeway driving beside a tank truck hauling a cylindrical trailer. We have noted diamond shaped placards with certain numbers and colors on the trailer. The trailer will have placards on both sides and the rear. The manifest will be on the front seat beside the driver. DOT requires this and for good reasons. Assume the tank truck had an accident and the trailer overturned and began leaking its contents onto the freeway. Lives might depend on knowing what the escaping liquid was and its hazards. The placards (see Figures 19-3 and 19-4) identify the contents by numerals on the placard and also tell the police if the cargo is flammable, corrosive, explosive, or a poison. When the fire department arrives, the placard will tell them the hazards they face in containment and cleanup, what PPE to wear, and what materials to use for cleanup. They get all this information by using the number on the placard to look up the chemical in an *Emergency Response Guidebook*.

Process technicians are responsible for compliance with certain DOT regulations because they will:

- Make preliminary physical inspections of cargo trailers and railcars before loading chemicals in them.
- Load cargo trailers/railcars with product.
- Placard or verify the correct placarding of the trailer/railcar.
- Inspect and sign the manifest.

Figure 19-3 Placarded Tank Truck

Figure 19-4 DOT Placards

Process technicians will receive training in shipping duties when they are hired. After receiving the training, the technician is responsible for complying with the DOT provisions that apply to their tasks.

UN or NA Identification Numbers

When hazardous materials are transported in tank cars, cargo tanks, portable tanks, or bulk packagings, United Nations (UN) or North American (NA) numbers must be displayed on placards, orange panels or, when authorized, plain white square-on-point configuration. UN or NA numbers are found in the Hazardous Materials Tables, Sections 172.101 and 172.102. UN numbers (see Figure 19-5) are used throughout the world. A cargo tank loaded and placarded with a UN number 2382 in the United States may be shipped to China. If the cargo vessel were to rupture while in China, Chinese emergency responders may not read or understand English, but they do understand the Arabic numeral system. They can locate that UN number in their translated Hazardous Materials Tables and identify the chemical as 1,2-Dimethylhydrazine and respond to the spill appropriately.

OTHER AGENCIES THAT AFFECT THE PROCESS INDUSTRY

Many other agencies affect the process industry and will vary from state to state. In Texas (the author's state), various agencies such as the Texas Commission on Environmental Quality, Railroad Commission, and General Land Office, all issue regulations that must be studied and complied with. Other organizations that affect the processing industry are Local Emergency Planning Commissions (LEPC), the Chemical Manufacturer's Association (CMA), and the American Petroleum Institute (API).

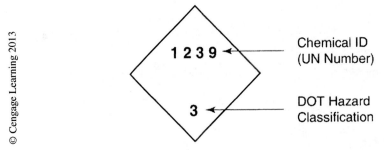

Figure 19-5   Explanation of Placard Numerals

SUMMARY

The Occupational Safety and Health Act was passed in 1970 because workplace accidents were causing an average of 14,000 deaths every year in the United States. Each year, 2.5 million workers were disabled in workplace accidents, and approximately 300,000 new cases of occupational diseases were reported. OSHA's mission is to ensure to the extent possible that every working person in the United States has a safe and healthy working environment.

OSHA developed standards based on its perception of need at the request of other federal agencies, state and local governments, and labor organizations. It uses the committee approach for developing standards. Before any of these actions can be undertaken, OSHA must publish its intentions in the *Federal Register.* Once a standard has been passed, it becomes effective on the date prescribed.

OSHA compliance officers can enter workplaces at reasonable times where work is taking place, inspect condition, facility machine, equipment, or materials, and question in private an employee or other person formally associated with the company. OSHA is empowered to issue citations and/or set penalties. Citations are issued for (a) other than serious violations, (b) willful violations, (c) repeat violations, and (d) failure to correct prior violations.

The DOT has as its primary function the ensuring of transportation safety and the protection of consumer interests. Process technicians are responsible for complying with certain DOT regulations because they will make preliminary physical inspections of cargo trailers and railcars before loading chemicals in them. They will placard or verify the correct placarding of the trailer/railcar, and inspect and sign the manifest.

REVIEW QUESTIONS

1. (Choose one) OSHA was created in 1970 with the mission of attempting to reduce:
   a. The number and severity of accidents by making equipment and procedures safer by mandatory means.
   b. The number and severity of accidents by making equipment and procedures safer by voluntary means.

2. (T/F) OSHA and its regulations are not enough to have a successful safety program.

3. (Choose two) The OSHA Act prescribes:
   a. Minimum safety and health standards
   b. Maximum safety and health standards
   c. Authorizes inspections

4. The three OSHA agencies are _____, _____, and _____.

5. A record of work-related illnesses and injuries is kept in the _____.

6. OSHA does not inspect businesses of _____ or fewer employees unless a fatality has occurred or there has been an employee complaint.

7. (T/F) OSHA has regulatory authority over coal mines.

8. (T/F) OSHA personnel may conduct workplace inspections unannounced.

9. Briefly describe how an OSHA workplace inspection would proceed from the first step to the last.

10. _____ are violations that have a direct relationship to job safety and health but probably would not cause death or serious harm.

11. _____ are violations that the employer intentionally and knowingly commits.

12. _____ are violations where, upon re-inspection, a substantially similar violation is found.

13. Employers must report the following types of accidents within _____ hours: (1) those that result in death, and (2) those result in the hospitalization of five or more employees.

14. List five employer responsibilities.

15. List five employee rights.

16. List five employee responsibilities.

17. Define the following legal terms as they relate to workplace safety: *negligence, willful/reckless conduct, care, contributory negligence.*

18. The purpose of _____ is to ensure transportation safety and protect consumer interests.

19. A _____ warns of the hazards within a vessel and/or the hazard faced in cleaning up a spill from the vessel.

20. (T/F) Process operators must inspect the running gear on railcars and tank trucks.

21. (Complete this sentence) The purpose of a UN number is to _____.

EXERCISES

1. Go to the OSHA website and make a list of 10 OSHA violations and the fines associated with each.

2. Many managers of businesses consider OSHA an intrusive nuisance and want the federal government to cancel all funding for the OSHA Administration and let it die. Write a one page report justifying whether you agree OSHA should be allowed to die, or take the opposite view and write a one page report justifying why you think OSHA should continue to function.

RESOURCES

www.wikipedia.org

www.osha.gov

www.YouTube.com

www.cdc.gov/niosh

CHAPTER 20

Environmental Protection Agency (EPA)

Learning Objectives

Upon completion of this chapter the student should be able to:

- *Describe the environmental responsibilities of the process employee.*

- *List the four primary functions of the Environmental Protection Agency (EPA).*

- *Explain the role analyzers play in environmental compliance.*

- *Describe how the EPA enforces environmental regulations.*

- *Explain why the Resource Recovery and Conservation Act (RCRA) was created.*

- *State why the Toxic Substances Control Act was created.*

INTRODUCTION

Process employees are responsible for protecting the environment both inside their production facility and outside it. They protect the environment outside their site by preventing incidents that result in a release of chemicals into the air, water, or soil beyond their site fence. They are also involved in protecting the public while cargo vehicles and vessels are transporting their products. Many operators are involved in preparing hazardous cargoes for transportation on public highways, railways, and waterways. They load, and in some instances placard tank trucks, rail cars, and barges. They inspect bills of laden and sign documents that, if filled out wrong, could lead to civil and/or criminal penalties.

For the last few decades, the people of the United States and their judicial system have placed a high priority on the protection of the environment. That protection includes special environments such as national forests, wilderness areas, or scenic rivers, and it also extends to such common place natural resources as the air we breathe, the water we drink, and the soil of our community. No country can allow the lakes and rivers that supply its drinking water to become contaminated with toxic chemicals, the air of its cities to become so unhealthy that its children become sickly, and its soil to become so polluted that crops grown on it cannot be consumed.

The U.S. Congress authorized the Environmental Protection Agency (EPA) to write regulations to protect the environment and over the last 30 years regulatory agencies and industry have collectively done a remarkable job at improving the environment and workplace. Some argue it was the regulatory agencies that should be credited with improvements in the environment and workplace. For whatever reason, there is a high level of environmental consciousness today. With the issue of the International Organization for Standards (ISO) 14000 standard for environmental management, the bottom line impact created by an effective environmental program has gained corporate attention. Thanks to hazardous and non-hazardous waste minimization and process improvements, profitability has increased and management can point to data that proves their facilities are more environmentally friendly. Of special concern to management is the fact that the legal system is now holding executives personally responsible for environmental pollution.

ENVIRONMENTAL PROTECTION AGENCY

The EPA was established in 1970 to protect the environment from pollution. Creation of the agency brought 15 federal programs under one management umbrella. Initially, the EPA focused on recycling and cleaning up open dump sites, but today the government has passed over 12 environmental laws that impact air, water, and land. These laws hold the manufacturers responsible for the hazardous wastes they generate from *cradle to grave*. This is a concise way of saying from the time the chemical is manufactured until it is properly disposed of.

The EPA's primary functions are to:

- Establish and enforce national standards for air and water quality.
- Establish and enforce national standards for air and water individual pollutants.
- Monitor and analyze the environment.
- Assist state and local governments with pollution control programs.

The EPA, like any other regulatory agency, sets standards (laws) and then checks to see that the laws are obeyed. It must enforce the Clean Air Act, the Clean Water Act, and Resource Recovery and Conservation Act (solid waste control). One way it does this is by monitoring industrial sites to determine if the site is deliberately releasing pollutants into the environment. Operators, instrumentation and analyzer technicians, and maintenance personnel working in refineries, petrochemical, and manufacturing plants are responsible for the proper operation of their equipment and proper performance of their duties so as to keep their facility in compliance with environmental standards. These employees will receive training on

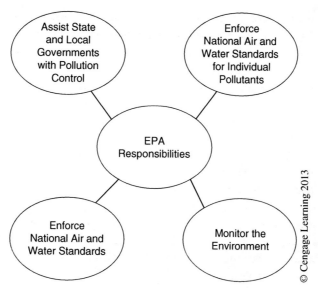

Figure 20-1 EPA Responsibilities

environmental standards where their jobs require it. The training makes the employee liable for environmental fines and penalties should they fail to perform their job correctly.

In the United States, the Occupational Safety and Health Administration (OSHA) is responsible for ensuring the safety of the environment within an industrial plant. Environmental problems caused by a facility that reaches beyond the fences of the facility become the concern of the EPA. Environmental problems can become lethal to human life if left uncontrolled. This is what happened in Donora, Pennsylvania in 1948 that resulted in the deaths of 20 people. Donora was situated in a low spot of the surrounding terrain. A meteorological temperature inversion blanketed the area for five days. Near Donora, a steel mill, sulfuric acid plant, and zinc smelter poured sulfur dioxide into the atmosphere. The weather inversion kept the sulfur dioxide close to the ground and people sickened and died.

The EPA was created with several environmental responsibilities (see Figure 20-1). The EPA enforces environmental regulations through a combination of civil and criminal penalties, including compliance and prohibition orders, fines, and imprisonment. Compliance orders require that persons, states, or businesses follow the requirements of the regulations. Prohibition orders require that persons, states, or businesses stop actions that violate regulations. Compliance and prohibition orders are the most commonly used enforcement tools. Penalties that result in fines can be very expensive, depending on the severity of noncompliance.

WATER POLLUTION CONTROL

The Clean Water Act of 1972 affects processing industries in at least two major ways. It requires (1) a company to have a water permit and it (2) regulates plant wastewater. The 1987 amendments directed states to develop and implement non-point source pollution management programs and to encourage them to pursue ground water protection activities.

The permit restricts a processing plant from taking all the water it wants to from nearby rivers or lakes. The plant has a permit to take only so much. Also, after using the water, it cannot return the water to a lake or river in just any condition. The water will have to be reasonably "clean" enough to meet compliance standards. This is a dramatic difference from the lack of standards in the 1960s that allowed one river to become so polluted it caught on fire!

Wastewater Standards

Under the Clean Water Act, all discharges into the nation's waters are unlawful unless specifically authorized. Industrial and municipal dischargers must obtain permits from the EPA or their municipal government before allowing any effluent to leave their premises. The Clean Water Act and its amendments have a large impact on most process industries and involve process technicians to catch and analyze samples and operate wastewater facilities; instrument technicians to maintain the instruments of the wastewater system; and analyzer technicians to maintain and calibrate analyzers on the wastewater system (see Figure 20-2) and outfalls.

Wastewater standards are applied to process wastewater, storm water, and once through water, cooling tower and boiler blowdown. Wastewater standards set a standard (quality) for wastewater before it can leave the plant and be returned to rivers, streams, or bays. Water sources leaving the processing site to return to the environment are called **outfalls**. Outfall samples are collected to be analyzed prior to releasing wastewater back into the environment to ensure environmental compliance. Some plants have analyzers mounted on outfall piping or vessels because of the critical nature of environmental compliance. This places a great responsibility on analyzer technicians for environmental compliance.

Process wastewater is water from process vessels and equipment that contains oils and chemicals. It also includes sanitary waste from bathrooms. Chemical spills on process pads washed into process sewers contribute significant amounts of chemicals and oils to

© Cengage Learning 2013

Figure 20-2 pH Analyzer

process wastewater. Storm water is rain that has fallen on the plant and its equipment and in the process accumulated oils and chemicals. It is now contaminated and is trapped in storm water ditches or storm water ponds. Before it can be released into the environment, it must meet environmental compliance. Once through cooling water, and cooling tower and boiler blowdown may contain large quantities of dissolved solids, some that may be hazardous in high concentrations. Also, heat exchangers may have leaks from the process side that contaminate the water. These waters, too, must meet environmental compliance before being released into the environment.

AIR POLLUTION CONTROL

The Clean Air Act, which has been amended many times, was created for several reasons, two of which are (1) develop a regional air pollution control program, and (2) enforce air quality standards. The Clean Air Act amendments of 1990 organized the Clean Air Act into nine separate titles, of which Titles I, III, IV, and V directly affect the process industries.

Air pollution occurs when the concentration of natural and/or man-made substances in the atmosphere becomes excessive and the air becomes toxic. Emissions from transportation, industry, and agriculture are man-made sources of air pollution. **Primary pollutants** are gases, liquids, and particulates dispersed into the atmosphere through either man-made or natural processes. In the United States, the primary pollutants are carbon monoxide, sulfur dioxide, nitrogen oxides, volatile organic compounds (VOCs), and particulate matter (soot, dust, etc.). **Secondary pollutants** are derived from primary pollutants that undergo a chemical reaction and become a different type of toxic material. In the United States, secondary pollutants are ozone, photochemical smog, and acid rain.

The refining and petrochemical industries are affected by the Clean Air Act because of the potential for unintended releases into the atmosphere. Flares inside plants are required to burn clean and smokeless. This is so critical that some plants have TV cameras that monitor the flares for clean burning. A flare that smokes too frequently or too long may lead to environmental fines. Scrubbers are required on furnace stacks to remove sulfur oxide gases which have the potential to be converted to sulfurous or sulfuric acid. Sulfur oxide gases released into the air can cause acid rain. It is not uncommon for a state environmental agency or EPA vehicle to be parked outside a plant monitoring for illegal emissions. Operators need to understand that the proper operation of their equipment serves several important functions besides safety and profits. It also involves the health of the environment and neighboring community, and the avoidance of civil or criminal penalties. Instrumentation and analyzer technicians also play a role in environmental compliance. Faulty instruments can lead to process upsets, spills, and releases.

Title V Permits

Operating permits are legally enforceable documents that permitting authorities issue to air pollution sources after the source has begun to operate. Most large sources and some smaller sources of air pollution are required to obtain a Title V permit which comes from Title V of the Clean Air Act as amended in 1990. Most title V permits are issued by state and local permitting authorities. Permits include pollution-control requirements from federal or state regulations that apply to a source. Some air quality compounds emitted by the processing industry and regulated by Title V are:

- Ground level ozone
- Particulate matter
- NH3-Ammonia
- Hg-Mercury
- Carbon monoxide
- Nitrogen oxides
- Sulfur dioxide
- Lead

The effects of the various pollutants include (1) the greenhouse effect from carbon dioxide, (2) acid rain from SO_x and NO_x, (3) ozone depletion from volatile organic compounds (VOCs), and (4) ground level health effects from ozone and particulates.

Permits contain specific requirements for facilities to operate pollution control equipment, monitor and limit pollution emissions, and report violations. Figure 20-3 is a TV in a control room receiving images from a camera aimed at a flare to verify that it still lit. The flare burns hydrocarbons and converts them into carbon dioxide and water vapor. If the flare went out, a variety of noxious hydrocarbons might be released into the surrounding community which could result into a large fine for the company.

The primary benefit to the public is that air permits limit the amount of air pollution allowed at a stationary source. For operating permits (Title V permits), a major source owner/operator is required to compile all applicable air pollution requirements at their source for purposes of obtaining one comprehensive permit (Title V permit). This process also includes public review of the proposed operating permit. Permits

© Cengage Learning 2013

Figure 20-3 TV Monitoring Flare

are enforceable documents. The public and the permitting authority may take action if a source fails to comply with its permit. The next paragraph details an example of a Title V violation and fine.

Murphy Oil USA Clean Air Act Settlement—On September 28, 2010, the U.S. Environmental Protection Agency (EPA) and the U.S. Justice Department announced that Murphy Oil USA has agreed to pay a $1.25 million civil penalty to resolve violations of the Clean Air Act at its petroleum refineries in Meraux, Louisiana and Superior, Wisconsin. As part of the settlement, the company will spend more than $142 million to install new and upgraded pollution reduction equipment at the refineries and also spend an additional $1.5 million on a supplemental environmental project.

Continuous Emissions Monitoring (CEMS)

Continuous emissions monitoring instruments (CEMS) are a key technology that helps monitor the emissions of pollution and dangerous substances into the atmosphere which can cause acid rain, global warming, or adverse health issues. Continuous emissions monitoring systems are required by some EPA regulations. A CEMS is defined as the equipment necessary to continuously monitor, analyze, and record the gas concentrations, particulate matter content, and/or flow rate of an emission non-stop day after day. They include the samplers, analyzers, software, and recording devices necessary to show compliance with EPA regulations.

The principal perceived drawback of continuous monitoring systems is cost: the conventional wisdom is that they are expensive to buy and expensive to operate. In fact, the purchase costs of CEMS have dropped considerably over the past decade, in some cases by over 50 percent, and continue to fall. The installation of CEMS (see Figure 20-4) will help industrial sources show good citizenship. Merely showing that industry is not hiding anything helps build good community relations. Further, good monitoring allows industry to show its environmental awareness.

© Cengage Learning 2013

Figure 20-4 Continuous Emissions Monitoring (CEMS) Analyzer

SOLID WASTE CONTROL

With the advent of the industrial revolution in the late 1800s to early 1900s, society was introduced to mass production. Clothing, automobiles, power tools, sports equipment, and furniture became affordable to the masses. An endless list of products made life easier. Arriving with this sudden plenitude was greater quantities of solid, liquid, and gaseous waste. Prior to mass production, waste was more a nuisance problem than a hazard. Mass production, industrialization, and creation of large quantities of organic compounds from crude oil saw huge volumes of waste generated that today would be classified as hazardous waste. In the early stages of the industrial revolution, no treatment, storage, and disposal facilities (TSDFs) existed to manage and destroy a company's hazardous waste. In fact, companies were not required to determine whether the waste generated at the end of the production line was hazardous. As a result, the common solution was to dig a hole in the ground and fill it with the waste. When that hole was filled, cap it off, and dig another one.

In the early 1970s, the Love Canal tragedy took the American public by surprise, which quickly changed to alarm. Love Canal was a housing development built over a capped industrial dumpsite in suburban Niagara Falls, New York. The population of this suburban area had a much higher than average incidence of certain types of cancer. Growing concern that such an event could "happen in our backyard" led to public demand for regulations that would ensure proper management of industrial waste. Two important federal regulations dealing with solid waste control are the Resource Conservation and Recovery Act (RCRA) and the Toxic Substances Control Act (TSCA).

Resource Conservation and Recovery Act (RCRA)

RCRA was designed to establish a national program to protect the nation's natural resources (air, water, and soil) from improper handling and storage of hazardous wastes. Congress charged the EPA with the task of tracking this regulation. The codified RCRA regulations can be found in 40 CFR Parts 240 through 282 (available on the Internet).

RCRA's main components include:

- Hazardous waste identification
- Cradle-to-grave manifest tracking
- Operating standards for generators and transporters of hazardous waste
- Permit system for TSDFs
- State authorization to assist in implementing the program

The three primary goals of RCRA are to (1) protect human health and the environment, (2) reduce waste and conserve energy and natural resources, and (3) reduce or eliminate the generation of hazardous waste as swiftly as possible. The three main players under RCRA regulation are (1) hazardous waste generators, (2) transporters, and (3) treatment, storage, and disposal facilities (TSDFs). The following discussion briefly examines each party's roles under RCRA.

Hazardous Waste Generator—The level of regulatory burden placed on a hazardous waste generator is determined by the amount of waste generated on a monthly basis. The three categories of generators are (1) conditionally exempt small quantity generators, (2) small quantity generators, and (3) fully regulated or large quantity generators.

Transporters—Hazardous waste transporters must comply with transporter regulations under the RCRA hazardous waste program (delivery, cleanup during transit, spill reporting, etc.), as well as licensing procedures for each state through which they will transport the waste. Transporters must accept waste only in accordance with the manifest system and comply with all U.S. Department of Transportation (DOT) regulations as they apply to hazardous waste shipments (e.g., driver qualifications, HazMat employee training, insurance). If a transporter stores hazardous waste at a transfer facility, the maximum duration of storage is ten days. The transporter must retain its copy of the hazardous waste manifest for a minimum of three years.

Treatment, Storage and Disposal Facilities—Any person who treats, stores, or disposes of hazardous waste is considered a TSDF. TSDFs must obtain approval to operate from the EPA or an authorized state agency.

Regulatory Enforcement

The EPA or an authorized state agency has three enforcement options under RCRA:

1. Administrative sanctions or penalties
2. Civil penalties
3. Criminal penalties

Administrative sanctions or penalties (non-judicial enforcement action) can be an informal action or an administrative order (e.g., compliance order, corrective action order). The maximum penalty is $25,000 per day of non-compliance for each violation. Civil penalties involve a formal lawsuit against a person who has failed to comply with some statutory/regulatory requirement or administrative order. The four types of civil penalties are: compliance action, corrective action, monitoring and analysis, and imminent hazard. The maximum penalty is the same $25,000 per day of non-compliance for each violation.

RCRA cites several acts that carry severe criminal penalties. They are knowingly:

1. Transporting waste to a non-permitted facility
2. Treating, storing, or disposing of waste without a permit
3. Omitting information and/or making false statement in any application, label, manifest, record, report, permit, or compliance document
4. Not complying with record keeping and reporting requirements
5. Transporting without a manifest
6. Exporting waste without the consent of the receiving country

The following generic properties or characteristics can cause a waste to be regulated as hazardous.

- Ignitability—liquids with a flash point of less than 140°F; non-liquid materials with the potential to spontaneously combust; DOT-ignitable compressed gases; and DOT oxidizers.

- Corrosivity—aqueous solutions with a pH of less than or equal to 2, or greater than or equal to 12.5; liquids that corrode steel at the rate of greater than 0.25 inch per year.
- Reactivity—normally unstable and undergoes violent change without detonating; reacts violently with water; forms potential explosive mixtures with water; produces toxic gases when mixed with water; and cyanide or sulfide containing matter.

Hazardous Waste Accumulation Areas

Hazardous wastes can be accumulated over a period of time at a processing site. Satellite accumulation is the accumulation of up to 55 gallons of a hazardous waste or one quart of acutely hazardous waste in containers at or near the point of generation. This waste is under the control of the operator of the process that generates the waste. The operator must keep hazardous waste containers in good condition, accumulate only with compatible materials, and the storage container is always closed during storage. Containers are labeled either "Hazardous Waste" or with other words that identify their contents (e.g., benzene waste). Once the 55-gallon container of hazardous waste or one quart of acutely hazardous waste has been accumulated, the generator must move the container to the facility's "less-than-90-day accumulation area" within three days and note the accumulation start date on the container label (see Figure 20-5).

Before 90 days is up hazardous waste is picked up for disposal. This will require a manifest. A manifest is the tracking document behind the entire RCRA "cradle-to-grave" concept. It is the official DOT shipping document that tracks a waste shipment from the time it is loaded on the transporter's vehicle until it is disposed (incinerated, fuel blended, landfilled, recycled, etc.). The manifest carries with it legal liability for the person who signs the document from the generating facility.

Figure 20-5 Hazardous Waste Accumulation Area

TOXIC SUBSTANCES CONTROL ACT

The Toxic Substances Control Act (TSCA) became law in 1976. It was intended to protect human health and the environment from the risk of chemicals in commerce. It resembles a consumer protection law to protect the public from new and existing chemicals anywhere in the community. It is designed to regulate commerce by requiring testing and necessary restrictions on certain chemical substances. TSCA has created a chemical inventory list which has over 70,000 chemicals.

The process employees working in industries that use, make, or transport chemicals, is required to receive training on TSCA. In addition, TSCA gives employees the right to report suspected toxic hazards to their company and requires the company to investigate such reports. Process industries must keep records of any adverse health effects or environmental effects of their chemicals. This data is reported to the EPA. Employee health records must be kept for 30 years and other records for five years.

ISO-14000 (A BRIEF DISCUSSION)

Another way the importance of the environment is being recognized is through ISO-14000. ISO-14000 is a series of standards and guidelines created by the International Standards Organization based in Geneva, Switzerland, designed to improve an organization's environmental management. It is not mandatory that companies embrace the standards and guidelines but if a company does become ISO-14000 certified there might be several advantages. If ISO-9000 certification history is taken as a guideline ISO-14000 may become necessary to do business on a global scale. Adapting ISO-14000 may improve a business's organizational performance, may generate marketing advantages, reduce environmental audits and penalties by regulatory agencies, may reduce product cost, and improve employee participation. ISO-14000 specifies requirements for establishing an environmental policy, planning environmental objectives and measureable targets, determining environmental aspects and impacts of products, activities and services, implementing and operating programs to meet objectives and targets, and performing a management review.

SUMMARY

The Environmental Protection Agency (EPA) was established in 1970 to protect the environment from pollution. Initially, the EPA focused on recycling and cleaning up open dump sites, but today the government has passed over 12 environmental laws that impact air, water, and land. These laws hold the manufacturers responsible for the hazardous wastes they generate from *cradle to grave*.

The Clean Water Act of 1972 affects processing industries in at least two major ways. It requires a company to have a water permit and it regulates plant wastewater. The Clean Air Act also affects processing industries. Control and disposal of hazardous waste became a function of the EPA through the Resource Recovery and Conservation Act. RCRA was designed to establish a national program to protect the nation's natural resources from improper handling and storage of hazardous wastes. Anyone who generates, transports, treats, stores, or disposes of hazardous waste must comply with RCRA. The Toxic Substances Control Act was designed to regulate commerce by requiring testing and necessary restrictions on certain chemical substances.

REVIEW QUESTIONS

1. (T/F) Process employees are responsible for protecting the environment inside their facility and outside it.

2. _____ is a term that means from the time the chemical is manufactured until it is properly disposed of.

3. _____ makes an employee liable for environmental fines and penalties should they fail to perform their job correctly.

4. _____ is responsible for ensuring the safety of the environment with an industrial plant.
   a. EPA
   b. DOT
   c. PSM
   d. OSHA

5. _____ orders require that persons, states, or businesses follow the requirements of the regulations.

6. _____ orders require that persons, states, or businesses stop actions that violate regulations.

7. List the four primary functions of the Environmental Protection Agency (EPA).

8. Two ways that the Clean Water Act affects the processing industries are by:
   a. Requiring a company to have a water permit.
   b. Requiring a company to pretreat the water it will use.
   c. Regulating plant wastewater.
   d. Inspecting wastewater facilities.

9. List three types of water that processing industries must cleanup before returning the water to the environment.

10. Water sources leaving the processing site to return to the environment are called _____.

11. _____ is rain that has fallen on the plant and its equipment and in the process accumulated oils and chemicals.

12. _____ are gases, liquids, and particulates dispersed into the atmosphere through either manmade or natural processes.

13. Two kinds of secondary pollutants are:
   a. Sulfur dioxide
   b. Ozone
   c. Volatile organic compounds
   d. Acid rain

14. CEMS is the acronym for _____.

15. _____ are legally enforceable documents that permitting authorities issue to air pollution sources.

16. List four air quality compounds regulated by Title V.

17. (T/F) A primary benefit to the public is that air permits limit the amount of air pollution allowed a stationary source.

18. The 1970 tragedy of _____ led to the regulation of industrial wastes.

19. Explain why the Resource Recovery and Conservation Act (RCRA) was created.

20. List the three primary goals of RCRA.

21. List three acts under RCRA that carry severe criminal penalties.

22. (T/F) TSCA has created a chemical inventory list which has over 70,000 chemicals.

23. Employee health records must be kept for _____ years.

EXERCISES

1. Go to the Environmental Protection Agency website and using their data, make a list of ten fines and the reasons for the fines.

2. Go to the Department of Transportation website and using their data make a list of ten fines and the reasons for the fines.

RESOURCES

www.wikipedia.org

www.epa.gov

www.YouTube.com

www.dot.gov

tceq.state.tx.us

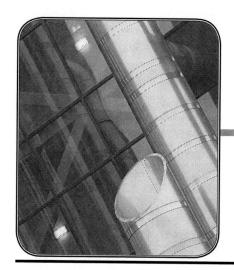

CHAPTER 21

Stress, Drugs, and Violence

Learning Objectives

Upon completion of this chapter the student should be able to:

- *Describe two types of violence in the workplace.*

- *Define hostile atmosphere.*

- *List three types of administrative controls to prevent violence.*

- *Describe four economic consequences to businesses caused by drug and alcohol abuse.*

- *Describe three physical and/or psychological functions affected by stress from shift work.*

- *List four sources of workplace stress.*

- *Describe five ways management can seek to reduce stress in the workplace.*

INTRODUCTION

It is unfortunate that the majority of highly developed countries have significant numbers of its citizens afflicted by "diseases" that are symptomatic of advanced societies. These diseases are violence, drug abuse, and stress. They are not confined to homes and neighborhoods, but too often appear in the workplace. Since human beings spend about one-third of their day at work, they have a high probability of eventually witnessing or being directly involved in an incident caused by one of the three "diseases." Violence,

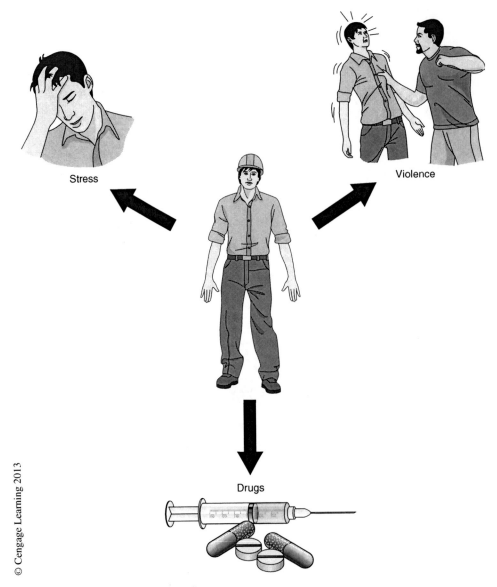

Figure 21-1   Violence, Drugs, Stress

drug abuse, and stress (see Figure 21-1) take a high toll in the workplace, not just of the workers but of the profits and productivity of a company. Although cardinal rules, policies, and employee assistance programs exist to control and/or eliminate violence, drug abuse, and stress, these "diseases" continue to persist. This chapter explores each of these human problems and how companies seek to eliminate and control them.

VIOLENCE IN THE WORKPLACE

The U.S. Department of Justice statistics on victims of crime, as found in the 1994 National Victimization Survey, indicates, excluding homicides, that during that year one million workers were victims of violence at work. In fact, from 1987 to 1993, approximately one million persons annually were assaulted at work. One in six violent crimes in the

United States occurs at work. Death and injury from workplace assaults can be cited under the OSHA Act of 1970. The employer is obligated to address workplace violence under the General Duty Clause. This clause, Section 5(a)(1), states that "each employer shall furnish to each his employees employment and place of employment which are free from recognized hazards that cause or are likely to cause death or serious physical harm to his employees."

Sometimes the violence is premeditated, as when a worker plans to assault a fellow worker. People are targeted because the aggressor believes the victim had something to do with his current situation (grievances or injustices). Most workplace incidents of violence are caused by stress that causes a worker to react with uncontrolled rage that results in violence. Many companies have cardinal rules that preclude fighting and acts of violence. Some also have policies against creating a "hostile atmosphere," which means intimidating fellow employees. Then there is the possibility of violence from non-employees. Examples would be rape or robbery in the company parking lot. To reduce violence and the opportunity for violence, employers must have an effective workplace violence prevention program that considers both internal and external sources of violence.

Violence Prevention Program

An effective workplace violence prevention program starts with the commitment of management to the program and the involvement of employees followed by a written workplace violence prevention policy statement. The policy should state that the employer refuses to tolerate violence in the workplace and is committed to the development and implementation of a program to reduce incidents of violence in the workplace.

The first step in controlling violence in the workplace is to assess the threat of violence. This could be accomplished with a team composed of management, human resources, labor, legal, etc. They will assess the potential for violence at the work site and recommend preventative actions. The assessment should include a common-sense workplace security analysis and review of relevant records. It will identify where and to whom the risk of violence exists by looking at processes and procedures, noting high-risk factors, and evaluating existing security measures. Employee surveys should solicit ideas on the potential for violent incidents and identify the need of security measures. Independent consultants such as law enforcement or security specialists can offer unbiased viewpoints to improve a violence prevention program.

The second step is to put a system in place that provides management with situation awareness through the tracking and analyzing of incidents and monitoring of trends. The third step in the program is to provide hazard prevention and control through engineering and administrative intervention. Engineering controls involve workplace design or redesign to correct problem areas. Physical barriers and security devices such as bulletproof enclosures, surveillance equipment, lighting, alarms, call boxes, two-way radios, and metal detectors are included in engineering controls. Administrative controls include, but are not limited to:

- Providing identification cards for employees
- Escort and badging policies for nonemployees
- Banning weapons in the workplace
- Employee training in situation awareness/avoidance/response

All employees should be trained in (1) techniques for recognizing the potential for violence, (2) conflict resolution, (3) how to react during incidents of violence, and (4) how to obtain assistance, including medical help. To help minimize the threat of violence in the workplace, many employers school their employees in conflict resolution which is a method of solving relational problems before they get out of control.

DRUGS IN THE WORKPLACE

In the past 30 years, the United States has experienced alarming level of drug use through all socioeconomic levels of its population. Today, drug abuse has intruded into the workplace and is profoundly affecting work safety. According to such sources as the Urban Institute and the House Select Committee on Narcotics Abuse and Control, drug-related absenteeism and medical expenses cost businesses about 3 percent of their payroll, and employers report that as many as one-in-five workers has a problem with alcohol or drug abuse. The consequences include an increase in absenteeism, health care claims, and lost worker productivity.

Drug Use by Employees

Drug use by employees in industry occurs among all levels of personnel, from top management to janitor. Drugs detract from work performance and safety. A significant number of industrial accidents were the direct result of using alcohol or other drugs on the job. Many drugs have long-term effects that last for days or even weeks, as is the case with marijuana and phencyclidine (PCP) (see Table 21-1). Others have after effects that can be as dangerous as the drugs themselves. Motivation for drug use in the workplace is varied. Some drug use is an attempt by the employee to work harder or to be more productive. Piece-workers may take amphetamines or opiates to enable them to work longer and faster without intolerable discomfort. Executives may use cocaine to allow them to work past their usual fatigue limit. Writers may use alcohol or cocaine to enhance their creative

Table 21-1 Drug Detection Time Table

Drug	Category	Detection Period
Amphetamine	Stimulant	2–4 days
Phenobarbital	Sedative/hypnotic	2–4 days
Cocaine	Stimulant	12–72 hours
Marijuana Casual use Chronic use	Euphoriant	2–days up to 30 days
Ethanol (alcohol)	Sedative/hypnotic	Very short
Opiates	Narcotic	2–4 days
Phencyclidine (PCP) Casual use Chronic use	Hallucinogens	2–7 days up to 30 days
Detection periods vary according to the metabolism of each drug user and should be viewed only as estimates.		

© Cengage Learning 2013

output. Many jobs have become a boring routine, plus the increasing automation of routine tasks means that an increasing number of employees are just standing and monitoring equipment. Monotonous, repetitive motion jobs and automation puts workers in situations that make them vulnerable to drug abuse.

As an individual's drug use progresses from casual to chronic use, the person who previously never used at work will begin to bring drugs to the workplace or come to work under the influence of drugs. Drug use at work may not occur until late in the addiction process—long after other personal functioning and relationships have been affected. The inevitable consequence of drug use in the workplace is the impairment of the worker's ability to perform, a loss of productivity, endangerment to themselves and their co-workers, and loss of employment.

Chronic drug users can experience serious effects of their drug problem, including (1) persistent brain dysfunction beyond the period of intoxication, (2) personal problems that preoccupy the employee's mental activities while at work, (3) and preoccupation with obtaining more drugs and anticipating their use. These effects interfere with and reduce the user's motivation to work. Employees that chronically abuse drugs can cause additional problems for employers through theft or misuse of company resources to purchase drugs. In addition, certain long-acting drugs like marijuana can produce impairment and increase error rate in complex tasks even when the user does not feel that they are still under the influence of the drug.

Behavioral Signs

Behavioral and psychological effects of alcohol and other drugs can include (1) frequently late to work, (2) absenteeism, (3) a decrease in performance and productivity, and (4) a change in mood or personality. Though there are definite symptoms of drug use, they may not always be present or easily identified. Sometimes the symptoms of drug use will resemble those of fatigue, a cold, or personal problems. The first adverse effect of substance abuse is often a behavioral or psychological change. A common indicator is the punctual employee who starts arriving late for work or calls in sick frequently. Another indicator is the employee's decline in productivity as they perform less efficiently and accurately. A third sign of trouble is borrowing money from other employees and failing to pay it back promptly.

The employee's mood can change as a result of drug abuse. They may develop new and persistent mood states such as depression, anxiety, anger, or paranoia. All drugs of abuse affect the user's mood, and these effects come and go according to the person's tolerance, the dose used, the frequency of use, and the duration of the drug's action. Most drug abusers have frequent unexplained mood changes. Certain drugs of abuse are associated with specific mood states or mental confusion. Cocaine and other stimulants produce paranoia and anger. The persistent use of depressants produces depression. Hallucinogens can produce lapses in concentration and memory and a range of serious psychiatric disorders.

Drug Testing in the WorkPlace

Companies drug screen potential employees before hiring them, then drug screen their workers at random. There is no schedule. It is almost like drawing names from a hat. It is not uncommon for a worker to be drug tested as often as three times a year. Usually, the worker is told by their supervisor they are wanted by the medical department and must report there within the next 30 minutes. At the medical department a sample (urine, blood, or hair) is taken for drug analysis.

The U.S. Department of Health and Human Services has issued a set of guidelines for drug testing in federal workplaces. These federal standards are used as a model for all urine testing programs. A urine sample is collected and an immunoassay is used for the initial screen, followed by a gas chromatography/mass spectrophotometer analysis (GC/MS). The urine must be tested twice to be confirmed and the second test must be of a different type from the first. Urine is collected under controlled conditions that ensure the correct identity of the sample. Then a chain of custody is created that demonstrates the sample is securely transported from the worksite to the testing laboratory, test results were accurately measured and recorded, and the findings accurately reported back to the worksite clinic.

Treatment

Employers are responsible for providing a safe work environment. If any employee is clearly unfit for duty, the company is liable for the safety of other employees as well as for the public good in the event of an accident. Management and the company's employee assistance program staff (EAP) must ensure that supervisory training stresses the need to intervene with an employee whose on-the-job behavior suggests physical or mental impairment or that the employee is a threat to self and/or others.

Different corporations have different standards of tolerance for drug use in the workplace. One corporation may have a "zero tolerance" standard, which means any detection of drugs and the employee is terminated. No excuses are accepted nor second chances and drug rehabilitation programs offered. Other companies allow employees one chance, require them to meet with an EAP personnel and enter a drug rehabilitation program. There is a growing trend in industry to use the positive drug screen as a method for early identification of workers with a drug problem and directing them to treatment. The employee is referred initially to the company's EAP and the EAP representative may refer the employee into counseling or some other form of treatment for chemical dependency. Essentially, it is the role of the EAP to evaluate the employee and to refer the employee into treatment. All of this is done in confidentiality. The employee may find themselves subject to more frequent drug testing by the company when they return to work.

STRESS IN THE WORKPLACE

We live in an era of constant change and demands for greater speed. In the workplace, management and workers face increasing demands caused by downsizing, layoffs, mergers, and pressure for continual improvement. Workplace stress is increasing and many people believe they are unable to do much to overcome its negative effects. Many workers will tell you they are now doing the job of several people since the company downsized. Many overloaded workers have a feeling of despair and suffer burnout. Working long hours takes a toll on personal relationships. According to NIOSH, the numbers of full- or part-time workers reporting high job stress shot up to 45 percent in 2002, up from 37 percent the year before. For many workers, the 40-hour work week has been a myth for years. Even when they are home, many workers are still tethered to the office by beepers, cell phones, and laptops.

Early signs of job **stress**, according to NIOSH, include short tempers, headaches, and low morale. Stress can lead to hypertension, heart attacks, and other diseases. Stress is the combination of adverse emotional and physical reactions people have to stressors (pressure, demands, and changes) in their environment. How they react can impair personal health and organizational effectiveness. Stress has been linked to almost every common disease, from heart disease to flu, and the rate of stress-related diseases is predicted to increase. Stressed workers are costly to an organization because stress affects productivity and innovation. The stress problem cuts both ways: organizational factors affect personal health and personal stress affects organizational health. Recognizing the costs of stress to individuals and organizations, many organizations have initiated stress-management programs that increase people's skills at managing stress and provide special help to people with stress problems.

Corporations tend to see stress as an individual problem due to an employee's lifestyle, psychological makeup, and personality. Workers view stress as result of excessive demands, poor supervision, or conflicting demands. However it is viewed, stress is a serious problem in the modern workplace. Stress-related medical bills and absentee rates cost employers about $150 billion annually (Smith, S.L. "Combating Stress," *Occupational Hazards*, March 1994). Workers who must take time off work because of stress, anxiety, or a related disorder will be off the job for about 20 days. Over $290 billion dollars is spent in the U.S. economy every year relating to compensation claims from on-the-job stress, health insurance, low-productivity, and disability (U.S. Bureau of Labor Statistics, 2010).

Stress is a reaction to various situations perceived by a person as threatening. Stress can be caused by (1) psychological factors, (2) social interactions between friends, co-workers, and family, (3) work, and (4) environment (temperature, pollution, etc.). Familiar causes of job stress are:

- Job security
- Deadlines
- Workload
- Supervisor and/or co-worker negativity
- Work place safety

Shift Work and Stress

Shift work is defined as work done primarily in other than normal daytime work hours. Shift work has traditionally been required by the medical community, the transportation industry, utilities, security, and, increasingly, by retail sales. Reduced alertness due to shift work has cost U.S. companies tens of billions of dollars a year due to lost productivity and safety problems. Companies implement shift work schedules based on productivity and efficiency. In humans, basic physiological functions are scheduled by the biological clock called the *circadian rhythm*. Most children in the United States grow up going to school during the day and sleeping at night. After a life of being accustomed to the day shift, the body notices a change in this rhythm. If the person takes a job starting at midnight, his or her body will still expect to be sleeping at night. Rotating shifts are the most stressful. Many physical and psychological functions are affected by circadian rhythm, such as blood pressure, heart rate, and body temperature. From a safety viewpoint, shift workers are subjected to more workplace stress in terms of weariness, irritability, depression, and a lack of interest in work.

Reducing Workplace Stress

Not all job-related stress can be eliminated, but people can learn to adapt to stress. Training can help people recognize and deal with stress effectively. Workplace stress can result from a perceived difference between the tasks demanded and a person's ability to cope with this demand. Under stress, the worker may develop feelings of tension, anger, fatigue, or anxiety. Even during good economic times workplace stress can be widespread. A few other sources of job stress would include the all too familiar (1) the company was recently bought out, (2) mandatory overtime is often required, (3) the consequences of making a mistake on the job are severe, and (4) few opportunities for advancement exist.

Reactions to Workplace Stress

Workplace stress may lead to decreased productivity, higher absenteeism, higher job turnover, poor morale, and stress-related illnesses. Some behavior patterns attributed to workplace stress are:

- Emotional stress, exhibited by anxiety, aggression, depression, etc.
- Performance stress, exhibited by being prone to accidents, emotional outbursts, etc.
- Physiological stress, exhibited by increased heart rate and blood pressure, heart disease, headaches, ulcers, etc.
- Organizational stress, exhibited by absenteeism, poor productivity, errors, high accident rates, burnout, etc.

Initially, the effects of stress may result in psychosomatic illness, but with continued stress, the symptoms eventually show up as an actual organic dysfunction. Continual or persistent stress has been linked to common physiological problems like colitis and gastric or duodenal ulcers. The harmful effects of stress can be reversed until the body's limit is reached. Stress continuing beyond the individual's limit results in disease. Psychosomatic diseases such as gastric ulcers, colitis, rashes, and autoimmune disorders may begin when the body becomes exhausted and can no longer adapt to stress.

Coping with Stress

Management can help design jobs in ways that lead to worker satisfaction and lessen work stress. In a highly competitive business environment, this may not be easy. Managers can assist by providing varied and independent work with occasions of collaboration with fellow workers and for personal development. Organizational approaches to coping with work stress include avoiding a monotonous, mechanically controlled work pace and mindless, constant repetition jobs.

Individuals also must learn to cope with stress. One of the most important factors in dealing with stress is learning to recognize its symptoms and taking them seriously. Keeping a positive mental attitude can help defuse some stressful situations. Also, workers can analyze stress-producing situations and decide whether it is worth the worry. There is no magic answer to workplace stress. In most jobs today, it will always be there in some form or other. Management must seek ways to reduce stress to increase worker health and productivity. Management should recognize workplace stress and take steps to reduce it in some of the following ways:

- Informing employees how to cope with stress.
- Employees are offered exercise and other stress reduction classes.

- Employees can work flexible hours.
- Workers are given the training, skills, and technology they need.
- Workers are given breaks and a place to relax during the workday.
- Employees are involved in making decisions that affect them.

SUMMARY

One in six violent crimes in the United States occurs at work. Death and injury from workplace assaults can be cited under the OSHA Act of 1970, which obligates the employer to address workplace violence under the General Duty Clause, which states that "employers shall furnish their employees employment and place of employment which are free from recognized hazards that can cause or are likely to cause death or serious physical harm to their employees." Most workplace incidents of violence are caused by stress that causes a worker to react with uncontrolled rage.

Workforce use of psychoactive drugs is costly to the employers because of increased injuries, lessened productivity, absenteeism, and more frequent mistakes on the job. Management and others in the workplace who are responsible for the safety and well-being of employees must be able to identify abusers who may be a danger to themselves and their co-workers. Observation of employees and their behavior is the first step in detecting a problem. This may be followed up with a variety of formal testing methods. Identifying, treating, and rehabilitating a chemically dependent employee is more cost-efficient than simply firing them and makes the most sense in terms of legal, labor relations, and other employer/employee concerns.

Stress is a pathological response to psychological, social, or occupational environmental stimuli. Workplace stress involves a worker's feelings resulting from the demands of the job and the person's capacity to cope with those demands. Sources of workplace stress include environmental conditions, work overload, role ambiguity, personality problems, and role conflict. Other sources of workplace stress are task complexity, job security, lack of psychological support, and safety concerns. Psychosomatic reaction to stress may eventually lead to disease. Workplace stress reduction can be accomplished by providing workers with information about how to deal with stress, talking with employees regularly, and permitting flexible work hours.

REVIEW QUESTIONS

1. Three "diseases" symptomatic of advanced societies are:
   a. Cardiovascular
   b. Stress
   c. Drug abuse
   d. Occupational hearing loss
   e. Violence

2. The employer is required to address workplace violence under OSHA's _____.

3. Most workplace incidents of violence are caused by _____ that causes a worker to react with uncontrolled rage.

4. A *hostile atmosphere* means:
   a. Threatening an employee with a gun or knife
   b. Pushing an employee
   c. Intimidating an employee

5. List three types of administrative controls to prevent violence.

6. About _____ workers has a problem with drug or alcohol abuse.
   a. One-in-four
   b. One-in-three
   c. One-in-five

7. The cost to businesses due to drug-related absenteeism and medical problems is 3 percent of their payroll. If the yearly payroll for a company was $2 million, the cost to the company is _____.

8. List three economic consequences to businesses caused by drug and alcohol abuse.

9. (T/F) All drugs of abuse affect the user's mood.

10. Three behavioral effects of drugs and alcohol on human beings are:
    a. Tardy to work
    b. Absenteeism
    c. Acting silly
    d. Drowsiness
    e. Change in mood or personality

11. (T/F) Companies drug test potential employees before hiring them and then at random after hiring them.

12. Early signs of job stress are _____, _____, and _____.

13. The definition of stress is:
    a. Unable to cope with a situation
    b. Situations that are physically intimidating
    c. A reaction to various situations perceived as threatening

14. List four sources of workplace stress.

15. Three psychosomatic diseases caused by stress are:
    a. Colitis
    b. Ulcers
    c. Brain cancer
    d. Rashes
    e. Diabetes

EXERCISES

1. Go to www.YouTube.com and view two videos on drug testing in the work place. Write a one page report on what you learned. Cite your sources.

2. Research the Internet and find a PowerPoint presentation on the stress created by shift work. Copy the presentation to a flash drive or CD and bring it to the class for presentation either by yourself or the instructor. Cite your source.

RESOURCES

www.wikipedia.org

www.osha.gov

www.YouTube.com

www.cdc.gov

www.cdc.gov/niosh

www.toolboxtopics.com

CHAPTER 22

Hurricanes, Plant Security

Learning Objectives

Upon completion of this chapter the student should be able to:

- *List four key elements of plant security and explain why the elements are considered crucial to plant security.*

- *Explain the importance of maintaining the integrity and effectiveness of operating controls.*

- *List three types of potential threats to the chemical and refining industries.*

- *Describe one type of prevention strategy*

- *Describe the key points of each phase in a hurricane standard operating procedure SOP.*

- *Explain the duties of a hurricane crew.*

INTRODUCTION

On September 11, 2001, the terrorist attacks on the World Trade Center Twin Towers and U.S. Pentagon ended America's illusion that the continental shores of the United States were immune from direct attack by terrorists. American society was forced to change in many significant ways, the most common of which is the extraordinary precautions taken to ensure the safety of airline passengers. More shocking news came out later when it was reported that terrorists could be planning attacks on chemical plants and oil refineries.

Commodity chemicals suddenly took on a sinister role as a potential terrorist tool as the processing industry became a potential terrorist target. Industry and society remembered the release of methyl isocyanate (MIC) at Bhopal, India, in 1994. This chemical release killed 3,800 people and injured approximately 11,000. Because of the availability and toxicity of toxic industrial materials (TIMs) like methyl isocyanate, these chemicals would be safer for terrorists to use to cause mass casualties than chemical or biological warfare agents. MIC is listed as a medium hazard on the TIMs Hazard Index List (see Table 22-1) created in 1998 by a North Atlantic Treaty Organization (NATO) international task force. The hazard index ranks TIMs according to the chemical's production, transport, storage, toxicity, and vapor pressure. Those chemicals ranked as high hazards are widely produced, transported, and stored, and they also have a high level of toxicity and vaporize easily. Approximately 850,000 U.S. businesses use, produce, or store TIMs, and many are located close to urban areas.

Chemical plants and oil refineries as well as the American Institute of Chemical Engineers (AICHE) center for process safety and the Chemical Manufacturer's Association (CMA) have developed documents to assist the chemical and oil refining industry develop site specific plant security guidelines. Some of the key elements of plant security include:

- Staffing
- Training
- Perimeter Monitoring
- Building Access
- Protection of the Internet and intranet information
- Regulatory Reporting
- Onsite computer and network access

Process employees are one of the key elements of site security (see Figure 22-1). They are on the frontline manning the foxholes against terrorists. Their knowledge, skills, and alertness are necessary for a successful "defense" of their production site. They know the faces of everyone who belongs on the unit, know if contractors are supposed to be on the unit, can check the fence line next to their unit, and notice anything out of the ordinary.

Table 22-1   Partial TIMS Hazard Index List

High Hazard	Medium Hazard	Low Hazard
Ammonia	Carbonyl sulfide	Arsenic trichloride
Chlorine	Methyl bromide	Bromine
Phosgene	Chloroacetone	Dimethyl sulfate
Sulfur dioxide	Phosphorous oxychlorine	Nitric oxide

© Cengage Learning 2013

Source: "U.S. Department of Justice. Guide for the Selection of Chemical Agent and Toxic Industrial material Detection Equipment for Emergency First Responders."

Figure 22-1   Key Elements of Site Security

THE SITE SECURITY PROGRAM

There are a variety of threats to the chemical and refining industries. Terrorists may attack a site with the intent to cause a catastrophic release or spill. Thieves looking for precursor chemicals to use in illegal drug manufacture may inadvertently cause chemical releases. Hackers hacking into information systems could corrupt computerized control systems and cause a chemical release. Unscrupulous hackers could create destructive conditions through modifications of fail-safe mechanisms.

To counter these threats, a good site security program would safeguard employees, environment, and community from chemical releases by maintaining the integrity and effectiveness of process operations. Control systems and sensitive equipment and information would be safeguarded. A site security program would reduce the costs of litigation and insurance and improve relationships with local authorities.

The first step in developing a security program is to conduct a risk assessment that determines what needs to be protected, the threats against those assets, and the consequences of an attack against those assets. Assets can be buildings, hazardous materials, pipelines, power lines, storage tanks, or process vessels. Two types of assessment are (1) a chemical hazards

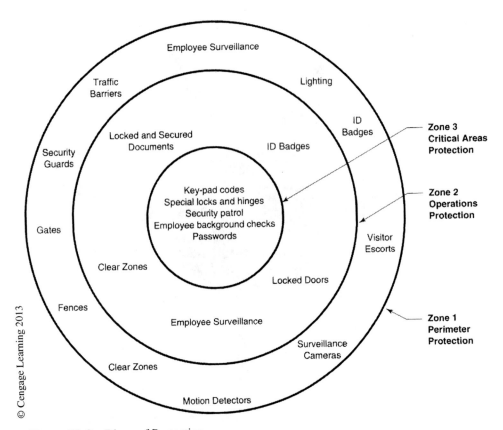

Figure 22-2   Rings of Protection

© Cengage Learning 2013

evaluation and (2) a process hazards analysis (PHA). Chemical hazards evaluations are routinely done in the chemical industry as part of the Responsible Care® Product Stewardship Code. It answers the question of how likely is a chemical release and how much harm would it cause. A process hazards survey is designed to highlight areas of potential vulnerability.

Prevention Strategies

One prevention strategy is *rings of protection* (see Figure 22-2). Physical security emphasizes rings of protection, meaning that the most important assets should be placed in the center of concentric levels of increasingly stringent security measures. For example, a site's process control center should never be placed near the entrance to a site. It should be protected in such a way that intruders would have to penetrate a fence at the property line (entrance gate), travel into the plant a significant distance without being detected, unlock the exterior door to the control building, and then use a keypad with a special code to gain entrance to the control room. Another prevention strategy involves networking with law enforcement agencies, security staff at other plants, and members of trade associations to share threat information. It is useful to know if other chemical sites have experienced an intrusion or vandalism so that appropriate security measures can be increased.

Incident Reporting and Analysis

Detailed records of security incidents allow managers to spot trends and collate facts. Incident management software has been developed that uses graphing and charting that can bring an offense or loss pattern to attention and identify issues of security concern. However, this requires managers to develop policies and procedures for incident reporting.

Employees serve as the eyes and ears of any company security effort. They know how equipment should be, the people who work specific areas, when contractors are scheduled to show up and for how long, what they should be doing, etc. In essence, they know the normal operations of a site and can spot when something abnormal is occurring. Training and awareness can transform employees into a natural surveillance system. Employees should be trained in a manner similar to the first responder in HAZWOPER. Their training should include such security practices as:

- Locking doors behind them
- Challenging strangers and/or people not wearing ID badges
- Not writing computer and access passwords near doors and computers
- Conducting plant perimeter surveillance near their unit when making their rounds
- Reporting suspicious persons or activities to plant security
- Avoiding leaving their computers on and unattended and to change their passwords frequently

Physical Security

The term *physical security* refers to equipment, building and grounds design, and security practices designed to prevent physical attacks against a facility's property, personnel, or information. It does not include cyber security. Physical security is accomplished by:

- Installing appropriate locks on exterior and interior doors
- Requiring visitor sign-in logs and escorts
- Scrutinizing access control at loading and unloading areas
- Installing appropriate penetration resistant doors and hinges
- Instituting a system of employee and contractor photo ID badges and train employees to challenge persons not wearing the badges
- Establishing a system that determines which vehicles (marine, rail, road, etc.) may enter the site, through which access point and under what conditions
- Establishing electronic access control for entry to motor control centers, rack rooms, server rooms, control rooms, etc
- Monitoring key areas of the facility via closed-circuit TV
- Employing motion sensors in sensitive areas

Perimeter Protection

Controlling the movement of people within a facility is important but it is much better to stop intruders at the edge of a facility's property before they reach vital assets and operational areas. Perimeter protection includes such measures as:

- Fences
- Traffic barriers that prevent vehicles from driving into the site from other than official entry points
- Personnel gates and turnstiles
- Clear zones and lighting that makes it possible to see intruders

- Security officers that tour a site and look for irregularities, staff site entrances, check IDs, etc.
- During periods of high terrorism alert, seek help from the local law enforcement authorities

Operations Security

Production facilities must protect information that could be useful to criminals and terrorists who intend to plan attacks on the site or steal hazardous materials for weapon building. Examples of site information that should be secured are:

- Plot plans
- Piping and instrumentation diagrams
- Process flow diagrams
- Formulations and recipes for products
- Process descriptions
- Emergency and nonroutine procedures

Managers should identify critical information and make it policy to shred documents instead of throwing them in the trash. They should identify file cabinets that should be kept locked and mark sensitive documents as "confidential." They should provide employee training and reminders about document security practices.

Computer and network security should be protected in several ways. Computer, server, and telecommunication rooms should be physically secured by locks. Computers should have firewalls and encryption available for sensitive documents. The principles of *least access, need to know,* and *separation of functions* should determine user authorizations rather than employee hierarchy. Only authorized personnel should have access to computer rooms and signs should not be posted indicating the location of the room.

Site security is every employee's concern, not just management's. Management must assess, develop a plan, implement the plan, and train its employees in security measures. The trained workforce must understand the plan and take it seriously. Employees cannot relax their guard after several months have gone by because nothing has happened at the site, in the region, or even nationally. Site security must take on the ever-vigilant urgency of site safety in fire and accident prevention. Only then will sites become secure.

HURRICANE PROTECTION

Natural disasters, such as hurricanes, floods, tornadoes, and earthquakes, happen. Plants must be prepared for them so that they suffer minimal damage to equipment and property or injury to personnel. Because hurricanes, unlike tornadoes and earthquakes, let us know when they are coming, we have the time to prepare properly for them. The hazards of hurricanes are:

- High winds
- Storm surge
- Rain and flooding
- Tornadoes
- Lightning
- Loss of services

Different plants may have different hurricane preparation procedures, depending on their location and nearness to water. What follows is a very generic hurricane preparation procedure that might be used by plants located near the Gulf of Mexico. Some sites divide the preparation, enduring, and cleanup sections of their hurricane standard operating procedures into phases.

Keep in mind that corporations do not blindly build multi-million dollar or billion dollar plants without consideration of the chances of natural disasters. They take into consideration prospective plant locations that would be subject to floods, hurricanes, and earthquakes. When building process units in a hurricane-prone area they study the frequency of hurricanes for that area and highest category hurricanes to strike. As an example, if a category 3 hurricane was the greatest storm to strike the area where a new plant is to be built, they will plan to build their plant to withstand a category 3 hurricane. Management intends for its investment to be safe and the people of the surrounding community to be safe from hazardous chemicals that might be released from their site by a storm.

Phase 1—A Storm Enters the Gulf

A plant prepares for a hurricane in a series of steps called phases. Phase I, as defined by the plant, consists of preparatory measures initiated when a potentially dangerous storm enters the Gulf of Mexico and is given a name by the weather service. All personnel are alerted via email or department meetings to the possibility of a hurricane. Hurricane crews, also called emergency crews, are alerted. Acquisition of supplies will begin within 48 hours of the estimated time of landing (ETL) of the hurricane. Teams throughout the plant, consisting of the first-line supervisor(s), team leaders, and upper management on site, will hold a brief meeting to determine if changes since the last hurricane alert will require modifying portions of their existing guidelines. During this phase, all loose objects around the unit will be tied down or stored inside appropriate facilities. All objects outside that might fly free in high winds will be tied down or secured in some manner. The plant will have to be staffed even if it is shutdown during the storm. Staffing will be by members of the hurricane crew, who are volunteers. Their job will be to safeguard plant property and equipment before, during, and after the storm until they are released.

Phase 2—Preparations

Activities during this phase begin 48 hours before winds in excess of 60 mph are predicted for the site area. Supplies for the hurricane crews, food, repair materials, cots, etc., will be received and distributed to the buildings to be occupied by the crews. Non-essential personnel may be released at this time. The initiation of shutting down certain process units in sequence begins. Unit teams will begin to closely monitor storm conditions. If best available information predicts a storm with winds less than 85 mph making landfall within 100 miles of the site, some process units may be taken offline. Staffing requirements for implementation of Phase 3 will be reviewed.

If a Phase 3 is likely, then within control rooms and other buildings where computers and file cabinets exist, personnel should begin to secure file cabinets and book cases from possible water damage, power down computers and printers, and cover with plastic. Depending upon the unit, the hazardous nature of its chemicals, the complexity and length of time to initiate shutdown, unit shutdown sequence may begin. Tanks, silos, and bins should be partially loaded with materials to provide additional weight to hold them in place should flood waters invade the plant. The buoyancy of huge empty tanks in several feet of water struck by high winds may cause them to shift or become deformed. Railroad cars and truck trailers that have not been removed from the plant must be secured. They should be filled with water to give them weight and brakes should be set and wheels blocked to keep them from rolling in the wind.

Because hurricanes can drop large quantities of rain in a short period of time, and also, if the plant is situated near the coast, an initial storm surge may lead to serious water problems. Storm sewers and drainage ditches on the plant site should be cleared of any restrictions or potential restrictions, and collection sumps should be emptied.

Hurricane crew members will be released to help their family members pack and escort them to the secure site (usually a major hotel away from the coast) plant management has selected. Hurricane crew members receive extra pay while performing hurricane duty. They are also given time to protect their personal property from hurricane damage. They should report back to the site within 16 hours ETL.

Phase 3—ETL 24 Hours

Phase 3 is sometimes referred to as the *abandonment phase*, meaning that plant units are shut down and all personnel released except for hurricane crew members. Orderly shutdown of all units will begin 24 hours before landfall of a storm with winds in excess of 85 mph. Unit supervisors should closely monitor those units that are shutting down. There will be only a maximum of 12 hours to accomplish all tasks associated with abandonment of the unit once Phase 3 is declared. During this phase, the unit shutdown sequence continues and food, water, and emergency supplies will continue to be distributed to the appropriate sites for hurricane crew members. The goal is to protect as much equipment as possible. There should be sufficient time allowed to prepare all necessary items prior to evacuation. All personnel not members of the hurricane crew will be released no later than 12 hours prior to landfall. All preparations should be completed as quickly as possible prior to release of non-essential personnel. Another inspection is made of process unit grounds for anything to be tied down or removed that might have been missed. Remaining personnel will ensure the emergency generator is fully fueled and unit vehicles are fully fueled and sheltered as much as possible.

Phase 4—ETL 16 Hours

There is not much left to be done during this phase. Hurricane crew members will report back to the plant and their assigned positions. Remaining operations personnel will be released as units are shut down and secured and hurricane crew members are available to the unit. At this time, all processing units will have been shut down unless management has determined some should remain up and circulating contents. Usually, the boilers remain in operation at this phase. If the boilers do not have to be shut down, then process units that have been shut down may be brought up much quicker because they have access to steam. This is management's call.

Phase 5—ETL 12 Hours

Hurricane crew members can only hunker down and wait for the storm. They will continue to monitor the storm's strength and path, and patrol the units and look for anything missed that should be secured. They will double-check all emergency equipment and verify they all have working radios. When the storm hits, they will monitor for damage, and when possible, make needed repairs.

Phase 6—Post Hurricane

Employees who were not members of the hurricane crew will be expected to return to work as soon as possible. Many employees may not be able to return right away for several reasons, such as (1) they left town to avoid the hurricane, (2) their homes were damaged and they must may them livable before returning to work, or (3) roads are impassable due to flooding and downed trees and debris. As fast as possible, the hurricane crew members will be released to check their homes and property and pick up their families. Employees may arrange for time off from work for personal business such as to repair or make arrangements for repair of their property. Cash advances are often available to help employees whose property suffered severe damage.

Since the main business of any business enterprise is to make money, the process units will be brought online as soon as safely possible. Personnel will begin to inspect all equipment for damage, inspect all exterior electrical junction boxes and control panels for moisture, and remove all flood and wind deposited debris. If the boilers and utilities were able to keep running during the storm, the units can be returned to service quickly. Depending upon the severity of the storm, units in the plant may be back up and lined out within a few days. A graphic summary of hurricane preparation is depicted in Figure 22-3.

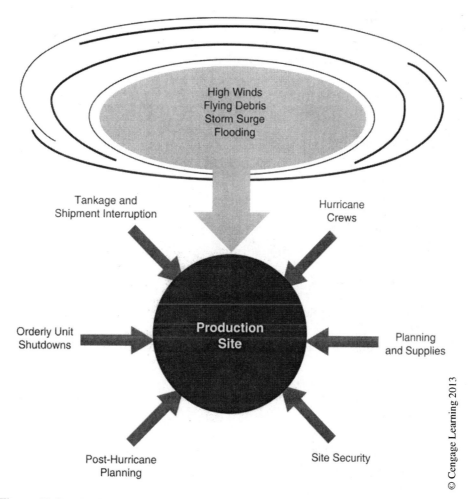

Figure 22-3   Hurricane Preparation

SUMMARY

The processing industry became a potential terrorist target after the attack on the World Trade Towers on September 11, 2001. Industry and society remembered the horrible human toll taken by the release of methyl isocyanate (MIC) at Bhopal. Approximately 850,000 U.S. businesses use, produce, or store toxic industrial chemicals (TIMs) and many are located close to urban areas. Process employees play a critical role in site security. Their knowledge, skills, and alertness are necessary for a successful "defense" of their production site. A good site security program would safeguard employees, the environment, and local communities from chemical releases by maintaining the integrity and effectiveness of process operations. Control systems and sensitive equipment and information should and must be safeguarded.

Hurricanes, unlike tornadoes and earthquakes, provide sufficient time to prepare properly for them. Different plants may have different hurricane preparation procedures, depending on their location and nearness to bodies of water. Some sites divide the preparation, enduring, and cleanup sections of their hurricane standard operating procedures into phases. The phases determine what tasks are to be accomplished during the time allotted for that phase, and by whom.

REVIEW QUESTIONS

1. _____ led to site security at processing industries becoming a critical safety issue.

2. _____ U.S. businesses produce, store, or transport toxic industrial materials?

3. Four key elements of plant security are:
   a. Training
   b. Two-way radios
   c. Staffing
   d. Chemical Manufacturer's Association
   e. Building access
   f. Perimeter monitoring

4. Describe two benefits of a site security program.

5. A chemical hazards evaluation reveals:
   a. How likely a chemical release is and how much harm it would cause
   b. The harmful cost of a chemical release to a company.

6. A process hazards survey:
   a. How likely a chemical release is and how much harm it would cause
   b. Highlights areas of potential vulnerability

7. Three types of potential threats to the chemical and refining industries are:
   a. Hackers hacking into information and control systems
   b. Fires and explosions
   c. Spills and releases

d. Attack by terrorists

e. Thieves looking for special chemicals

8. _____ emphasizes that the most important assets should be placed be placed in the center of concentric levels of increasing security measures.

9. (T/F) A prevention strategy could be networking with law enforcement agencies, security staff at other plants, and sharing information with other plants.

10. Explain the importance of process employees to site security.

11. Four types of physical security a plant can install or institute are:
   a. Appropriate locks on interior and exterior doors
   b. Requiring visitor sign-in logs and escorts
   c. Monitoring key areas with via closed-circuit TV
   d. Using motion sensors in key areas
   e. Stepping outside at intervals to look around
   f. All of the above

12. The term physical security means:
   a. Equipment, building design, and security practices
   b. Cyber security
   c. Both a and b

13. Three hazards of hurricanes are _____, _____, and _____.

14. Phase 3 is sometimes referred to as the _____.

15. Describe the key points of Phase 2 in a hurricane plan.

16. Describe the key points of Phase 3 in a hurricane plan.

17. Three duties of the hurricane crew are:
   a. Staff the plant during the hurricane
   b. Patrol the plant and look for anything missed that should be tied down
   c. Take their families to a designated secure place
   d. Monitor for storm damage and, if possible, repair storm damage
   e. All of the above

EXERCISES

1. Search the Internet and find information about damage sustained by refineries and petrochemical plants by natural disasters and write a one page report on what you discovered. Cite your sources.

2. Search the Internet and www.YouTube.com and find information about plant security today, what devices are used, information sharing, etc. Write a one page report on what you discovered. Cite your sources.

RESOURCES

www.wikipedia.org

www.osha.gov

www.YouTube.com

www.toolboxtopics.com

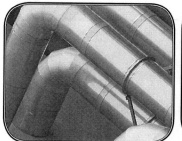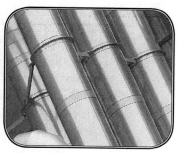

Glossary

A

Abatement Period—The amount of time given an employer to correct a hazardous condition that has been cited.

Absorption—Passage through the skin and into the bloodstream.

Accident Prevention—The act of preventing a happening that may cause loss or injury to a person.

Accident Rate—A fixed ratio between the number of employees in the workforce and the amount that are injured or killed every year.

Accident Report—Records the findings of an accident investigation, the cause or causes of an accident, and recommendations for corrective action.

Accident—An unplanned disruption of normal activity resulting in an injury, equipment damage, or a material release.

Acclimatization—Process by which the body becomes gradually accustomed to heat or cold in a work setting.

Acute Effect—An adverse effect on the body with severe symptoms which develop rapidly and that may subside after the exposure ceases.

Administrative Activities—Are the programs put into action, such as drills, exercises, audits, etc.

Administrative Controls—Procedures, policies, and plans that are adopted to limit employee exposure to hazardous conditions.

Administrative Programs—Are the written documents that explain how hazards are to be controlled.

Aerosols—Liquid or solid particles so small that they can remain suspended in air long enough to be transported over a distance.

Affected Employee—An employee whose job requires them to operate or use a machine or equipment on which servicing or maintenance is being performed under lockout or tagout, or whose job requires him/her to work an area in which such servicing or maintenance is being performed.

Anesthetics—A substance that can inhibit the normal operation of the central nervous system without causing serious or irreversible effects.

Asphyxiant—There are two types of asphyxiants: simple and chemical. Simple asphyxiants are gases that replace oxygen in the air and can cause death by suffocation. Some common asphyxiant gases are nitrogen and carbon dioxide. A chemical asphyxiant, such as carbon monoxide, prevents hemoglobin from adsorbing oxygen.

Assumption of Risk—Based on the theory that people who accept a job assume the risks that go with it. It says employees who work voluntarily should accept the consequences of their actions on the job rather than blaming the employer.

Audiogram—The results of an audiometric test to determine the noise threshold at which a subject responds to different test frequencies.

Audiometric Testing—Measures the hearing threshold of employees.

Auto-Ignition Temperature—The lowest temperature at which a vapor-producing substance or a flammable gas will ignite even without the presence of a spark or a flame.

Authorized employee—A person who locks out or tags out machine or equipment in order to perform servicing or maintenance on the machine or equipment.

B

Biological Hazards—Harmful molds, fungi, bacteria, and insects.

Bonding—Used to connect two pieces of equipment by a conductor, usually a copper cable. Also, involves eliminating the difference in static charge potential between materials.

Breakthrough—Occurs when the respirator cartridge becomes saturated with the contaminant and a small amount of the contaminant begins to enter the facepiece.

C

Carcinogen—Any substance that can cause a malignant tumor or a neoplastic growth.

Ceiling—The level of exposure that should not be exceeded at any time for any reason.

Chemical Hazards—Include mists, vapors, gases, dusts, and fumes.

Chronic Effect—An adverse effect on a human or animal with symptoms that develop slowly over a long period of time or which recur frequently.

Circadian Rhythm—Biological clock.

Code—A set of standards, rules, or regulations relating to a specific area.

Combustion—Is the process by which fire converts fuel and oxygen into energy, usually in the form of heat.

Combustion Point—The temperature at which a given fuel can burst into flame.

Combustible Substance—Any substance with a flash point of 100°F or higher.

Confined Space—An area with limited means of egress that is large enough for a person to fit into, but is not designed for occupancy.

Contamination—Occurs when a hazardous substance remains on the clothing, hair, skin, or other part of a person.

Contamination Reduction Zone (CRZ)—Surrounds the exclusion zone and is a rest area for workers and also serves as the staging area for the emergency response equipment. Typically, the CRZ is where decontamination of personnel and equipment is carried out.

Contributory Negligence—An injured worker's own negligence contributed to the accident. If the actions of employees contributed to their own injuries, the employer is absolved of any liability.

Corrosive—A substance that destroys human tissue as a result of direct physical contact. Typically, corrosives are acids and bases.

Critical burns—Second-degree bums covering more than 30 percent of the body and third-degree burns covering over ten percent.

D

Death rates—A fixed ratio between the number of employees in the workforce and the number that are killed each year.

Decibel—The unit applied when measuring sound. One-tenth of a bel. One decibel is the smallest difference in the level of sound that can be perceived by the human ear.

Degradation—Is the gradual chemical destruction of a material.

Dose—In terms of monitoring exposure levels, the dose is determined by how much of a substance a person is exposed to (called the concentration) and how long the exposure lasts (the time).

Dose Threshold—The minimum dose required to produce a measurable effect.

DOT—U.S. Department of Transportation; the DOT has specific requirements for the labeling of chemical hazards transported on public highways and waterways.

Dust—Is formed when solid material is broken down, such as in crushing and grinding operations.

E

Electrical Hazards—Potentially dangerous situations related to electricity (e.g., a bare wire).

Electrolytes—Minerals that are needed for the body to maintain the proper metabolism and for cells to produce energy.

Emergency Action Plan (EAP)—A plan for an anticipated emergency (e.g., fire, hurricane, chemical spill, and so on).

Emergency Response—The loss of containment for a chemical or the potential for loss of containment that results in an emergency situation requiring an immediate response.

Emergency Response Plan—A written document that identifies the different personnel/ groups that respond to various types of emergencies and, in each case, who is in charge.

Emergency Response Team—A special team that responds to general and localized emergencies to facilitate personnel evacuation and safety, shut down building services and utilities as needed, work with responding civil authorities, protect and salvage company property, and evaluate areas for safe reentry.

Employer-Biased Law—A collection of laws that favored employers over employees in establishing a responsibility for workplace safety.

Engulfment—The surrounding and effective capture of a person by a liquid or finely divided (flowable) solid substance that can be aspirated (breathed in) to cause death by filling or plugging the respiratory system or that can exert enough force on the body to cause death by strangulation, constriction, or crushing.

Ergonomic Hazards—Workplace hazards related to the design and condition of the workplace. For example, a workstation that requires constant overhead work is an ergonomic hazard.

Error of Commission—Performing a function not required, such as unnecessarily repeating a procedural step, adding unnecessary steps to a sequence, or doing an erroneous step.

Error of Omission—Failure to perform a required function.

Exclusion Zone—The contaminated area where site clean-up, drum movement, drum staging, and clean up are some of the activities conducted at the exclusion zone.

Explosion—A very rapid, contained fire.

Explosive Range—Defines the concentrations of a vapor or gas in air that can ignite from a source.

Exposure—Occurs when a chemical, infectious material, or other agent enters or is in direct contact with the body.

Exposure Threshold—A specified limit on the concentration of selected chemicals. Exposure to these chemicals that exceeds the threshold is considered hazardous.

F

Fall Arrest System—A system employed to protect a worker when the worker is at risk of falling from an elevated position.

Fellow Servant Rule—Employers are not liable for workplace injuries that result from negligence of other employees.

Fire—A chemical reaction between oxygen and a combustible fuel.

Fire Point—Is the minimum temperature at which the vapors of a substance will continue to burn given a source of ignition.

Fire Watch—A person responsible for surveying an area where work is occurring who will stop the work if a condition is detected that might lead to a fire and who sounds the alarm if a fire is detected.

Fire Hazards—Conditions that favor the ignition and spread of fire.

Fire Point—The minimum temperature at which the vapors or gas in air can ignite from a source of ignition.

First-Degree Burn—A mild inflammation of the skin known as erythema.

Flame—Resistant Clothing-Special clothing made of materials or coated with materials that are able to resist heat and flames.

Flammable Substance—Any substance with a flash point below 100°F and a vapor pressure of less than 40 pounds per square inch at 100°F.

Flash Point—The lowest temperature for a given fuel at which vapors are produced in sufficient concentrations to flash in the presence of a source of ignition.

Foreseeability—Concept that a person can be held liable for actions that result in damages or injury only when risks could have been reasonably foreseen.

Frostbite—Occurs when there is freezing of the fluids around the cells of the outer body tissues.

Frostnip—Less severe than frostbite, it causes the skin to turn white and typically occurs on the face and other exposed parts of the body.

Fugitive Emissions—Volatile organic compounds and other designated liquids or gases escaping from valves, piping, and equipment in minute amounts from each source but that cumulatively contribute to significant air pollution.

Fumes—Are fine particles formed when a volatized solid, usually a metal, condenses in air. They can be caused by welding and smelting operations.

G

Gases—Are substances that become airborne at room temperature.

Good Housekeeping—Proper cleaning and maintenance of a work area.

Ground Fault—When the current flow in the hot wire is greater than the current in the neutral wire.

Ground Fault Circuit Interrupter—Can detect the flow of current to the ground and open the circuit, thereby interrupting the flow of current.

H

Hazard—A condition with the potential of causing injury to personnel, damage to equipment or structures, loss of material, or lessening of the ability to perform a prescribed function.

Hazard Analysis—A systematic process for identifying hazards and recommending corrective action.

Hazard Communication Standard (HAZCOM)—An OSHA standard that addresses the assessment and communication of chemical hazards in the workplace.

Hazardous Material Identification System (HMIS) label—A color coded labeling system that warns of the hazards associated with a particular chemical.

HAZOP—Hazards and Operability Study.

Hazard And Operability Review—An analysis method that was developed for use with new processes in the chemical industry.

Hazardous Condition—A condition that exposes a person to risks.

Hazardous Waste Operations and Emergency Response (HAZWOPER)—An OSHA standard that addresses chemical spills and releases.

Health Hazards—Are defined as materials for which there is scientific evidence demonstrating that acute or chronic health effects may occur in exposed employees.

Hearing Conservation—Systematic procedures designed to reduce the potential for hearing loss in the workspace. Employers are required by OSHA to implement hearing conservation procedures in settings where the noise level exceeds a time-weighted average of 85dBA.

Heat Cramps—A type of heat stress that occurs as a result of salt and potassium depletion.

Heat Exhaustion—A type of heat stress that occurs as a result of water and/or salt depletion.

Heat Rash—A type of heat stress that manifests itself as small raised bumps or blisters that cover a portion of the body and give off a prickly sensation that can cause discomfort.

Heat Stroke—A type of heat stress that occurs as a result of a rapid rise in the body's core temperature.

Hepatoxin—A chemical that can produce liver damage in humans. Examples include carbon tetrachloride and nitrosamines.

Human Factor—Attributes accidents to a chain of events ultimately caused by human error.

Hypothermia—The condition that results when the body's core temperature drops to dangerously low levels.

I

Ignition Temperature—The temperature at which a given fuel can burst into flame.

Incident Commander (IC)—The leader that directs the emergency response actions.

Immediately Dangerous to Life and Health (IDLH)—Exposure to airborne contaminants that is "likely to cause death or immediate or delayed permanent adverse health effects or prevent escape from such an environment." Examples include smoke or other poisonous gases at sufficiently high concentrations.

Impact Accidents—Involve a worker being struck by or against an object.

Indirect Costs—Costs that are not directly identifiable with workplace accidents.

Industrial Hygiene—An area of specialization in the field of industrial safety and health that is concerned with predicting, recognizing, assessing, controlling, and preventing environmental stressors in the workplace that can cause sickness or other forms of impaired health.

Ingestion—Entry through the mouth.

Inhalation—Taking gases, vapors, dust, smoke, fumes, aerosols, and/or mists into the body by breathing in.

Interlocked Guards—Shut down the machine when the guard is not securely in place or is disengaged.

Interlocks—Automatically break a circuit when an unsafe situation is detected.

Irritants—Substances that cause irritation to the skin, eyes, and the inner lining of the nose, mouth, throat, and upper respiratory tract.

J

Job Descriptions—Written specifications that describe the tasks, duties, reporting requirements, and qualifications for a given job.

Job Safety Analysis—A process through which all of the various steps in a job are identified and listed in order.

K

Kinetic Energy—The energy resulting from a moving object.

L

Learning Objectives—Specific statements of what the learner should know or be able to do as a result of completing the lesson.

Levels of Protection—The four categories of equipment that protects the body against contact with known or anticipated toxic chemicals according to the degree of protection afforded:

Liability—A duty to compensate as a result of being held responsible for an act or omission.

Lockout/Tagout System—A system for incapacitating a machine until it can be made safe to operate. "Lockout" means physically locking up the machine so that it cannot be used without removing the lock. "Tagout" means applying a tag that orders employees not to operate the machine in question.

Lost Time—The amount of time that an employee was unable to work due to an injury.

M

Malpractice—Negligent or improper practice.

Material Safety Data Sheet (MSDS)—A document that contains all of the relevant information needed concerning specific hazardous materials.

Mechanical Hazards—Those associated with power-driven machines, whether automated or manually operated.

Mechanical Injuries—Injuries that have occurred due to misuse of a power-driven machine.

Micro-insult—A term that means exposure to small or supposedly insignificant amounts of a harmful chemical.

Minor Burns—All first-degree bums are considered minor as well as second-degree bums covering less than 15 percent of the body.

Mists—Tiny liquid droplets suspended in air, usually caused by spraying and cleaning operations.

Moderate Burns—Second-degree bums covering less than 30 percent of the body and third-degree bums covering less than ten percent are considered moderate.

Mutagen—A substance or agent that can alter the genetic make-up of a sperm or egg cell. Examples include ozone and radiation.

N

National Fire Protection Association's (NFPA) label—Has a diamond shape and uses the same color code as the HMIS label, but has different and more specific meanings for the numbers in each hazard category.

National Institute of Occupational Safety and Health (NIOSH)—Part of the Centers for Disease Control of the Department of Health and Human Services. It is required to publish annually a comprehensive list of all known toxic substances. It will also provide onsite tests of potentially toxic substances so that companies know what they are handling and what precautions to take.

Negligence—Failure to take reasonable care or failure to perform duties in ways that prevent harm to humans or damage to property.

Negligent Manufacture—The maker of a product can be held liable for its performance from a safety and health perspective.

Nephrotoxins—Chemicals which produce kidney damage in humans. Examples include uranium and halogenated hydrocarbons that are contained in many solvents.

Neurotoxins—Chemicals which have toxic effects on the nervous system. Examples include mercury and carbon disulfide.

Noise—Unwanted sound.

O

Occupational Diseases—Pathological conditions brought about by workplace conditions or factors.

Occupational Safety and Health Administration (OSHA)—The administrative arm for the Occupational Safety and Health Act responsible for regulating and enforcing safety and health policies for the United States.

The Occupational Safety and Health Review Commission (OSHRC)—An independent board whose members are appointed by the president and given quasi-judicial authority to handle contested OSHA citations.

Odor Threshold—The lowest concentration of a substance that a person can detect by smell.

Olfractory—Refers to the sense of smell.

Organized Labor—A group of employees who joined together to fight for the rights of all employees (i.e., unions).

Overexertion—The result of employees working beyond their physical limits.

P

Particulate Matter—A suspension of fine solid or liquid particles in the air, such as dust, fog, fumes, mist, smoke, or sprays.

Permissible Exposure Limit (PEL)—An exposure limit set by OSHA that is required by law.

Permeation—Is the amount of chemical that will pass through a material in a given area in a given time.

Personal Monitoring Devices—Devices worn or carried by an individual to measure chemical exposure and/or radiation doses received.

Personal Protective Equipment (PPE)—Any type of clothing or device that puts a barrier between the worker and the hazard (e.g., safety goggles, gloves, boats, hard hats, and so on).

Physical Hazards—These are hazards due to the physical properties of chemicals. They include extremes of temperature, compressed gases, explosives, flammables, and excessive radiation. Other types of physical hazards are noise and vibration.

Predictable errors—Are those which experience has shown will occur again if the same conditions exist.

Preliminary Hazard Analysis—Conducted to identify potential hazards and prioritize them according to (1) the likelihood of an accident or injury being caused by the hazard; and (2) the severity of injury, illness, and/or property damage that could result if the hazard caused an accident.

Primary Pollutants—Are gases, liquids, and particulates dispersed into the atmosphere through either man-made or natural processes.

Pressure—The force exerted against an opposing fluid or thrust distributed over a surface.

Productivity—The concept of comparing output of goods or services to the input of resources needed to produce or deliver them.

Q

Quality—A measure of the extent to which a product or service meets or exceeds customer expectations.

R

Random Errors—Are non-predictable and cannot be attributed to a specific cause.

Reactivity—The tendency of a substance to react and undergo chemical change. The products of these reactions may be hazardous such as fire, explosion, or the release of toxic gas.

Reasonable Risk—Exists when consumers (1) understand risk, (2) evaluate the level of risk, (3) know how to deal with the risk, and (4) accept the risk based on reasonable risk/benefit considerations.

Reclamation—A process whereby potentially hazardous materials are extracted from the byproducts of a process.

Repeat Violation—A violation of any standard, regulation, rule, or order where, upon re-inspection, a substantially similar violation is found.

Repetitive Motion—Short-cycle motion that is repeated continually.

Repetitive Strain Injury—A broad and generic term that encompasses a variety of injuries resulting from cumulative trauma to the soft tissues of the body.

Respiratory Protection—Consists of air cleaning or air supplying devices that protect your breathing system from contaminants or supply fresh air in toxic/oxygen deficient atmospheres.

Risk—A possibility of loss or injury.

Risk Analysis—An analytical methodology normally associated with insurance and investments.

S

Safeguarding—Machine safeguarding was designed to minimize the risk of accidents of machine-operator contact.

Safety Policy—A written description of an organization's commitment to maintaining a safe and healthy workplace.

Safety Trip Devices—Include trip wires, trip rods, and body bars that stop the machine when tripped.

Second-Degree Burn—A burn that results in blisters forming on the skin.

Secondary Pollutants—Are derived from primary pollutants that undergo a chemical reaction and become a different type of toxic material.

Sensitizer—A substance that can cause an allergic reaction in man that usually occurs only after repeated exposure to the chemical. The most common problem with this type of reaction is skin sensitization.

Shock—A depression of the nervous system.

Short-Term Exposure Limit (STEL)—The maximum concentration of a given substance to which employees may be safely exposed for up to 15 minutes without suffering irritation, chronic or irreversible tissue change, or narcosis to a degree sufficient to increase the potential for accidental injury, impair the likelihood of self-rescue, or reduce work efficiency.

Sprain—The result of torn ligaments.

Static Electricity—Electricity created by a surplus or deficiency of electrons on the surface of a material.

Strain—The result of over-stretched or torn muscles.

Stress—A pathological human reaction to psychological, social, occupational, or environmental stimuli.

Superfund Amendments and Reauthorization Acts (SARA)—Was designed to allow individuals to obtain information about hazardous chemicals in their community so that they are able to protect themselves in case of an emergency.

Support Zone—Contains people not directly involved in eliminating or controlling the emergency. This zone contains administrative personnel, a communication station, food and drinks, and other accessories.

Systemic Poison—A chemical that has a toxic effect within the body on one or more of the organs or bodily systems. Lead is a systemic poison of the nervous system.

T

Temporary Variance—Employers may ask for this when they are unable to comply with a new standard but may be able to if given time.

Teratogen—A substance that can cause birth defects in the fetus of a pregnant female. Examples include nicotine and alcohol.

Thermal Expansion Detectors—Use a heat-sensitive metal link that melts at a predetermined temperature to make contact and ultimately sound an alarm.

Tie-off point—Point where the lanyard or lifeline is attached to a structural support.

Third Degree Burn—A burn that penetrates through both the epidermis and the dermis. They may be fatal.

Threshold Limit Values (TLVs)—The level of exposure below which all employees may be repeatedly exposed to specified concentrations of airborne substances without fear of adverse effects. Exposure beyond the TLV is considered hazardous.

Time Weighted Average—The level of exposure to a toxic substance to which a worker can be repeatedly exposed on a daily basis without suffering harmful effects.

TLV—Threshold Limit Value. The airborne concentration of a substance that an average person can repeatedly be exposed to without any adverse effect. TLV's may be expressed three ways:

TLV-TWA—Time Weighted Average based on an allowable exposure averaged over a normal eight-hour workday or 40-hour work week.

TLV-STEL—Short Term Exposure Limit or maximum concentration for a continuous 15 minute exposure period not to exceed four such exposures per day.

TLV-C—Ceiling Exposure Limit or maximum exposure concentration not to be exceeded under any circumstances.

Toxic—A substance that is harmful to human health. Poisonous.

Trenchfoot—A cold weather condition of the foot that manifests itself as tingling, itching, swelling, and pain.

Two—Hand Controls-Requires the operator to use both hands concurrently to activate the machine.

U

Ultraviolet Detectors—Sound an alarm when the radiation from fire flames are detected.

Underwriters Laboratory (UL)—Determines whether equipment and materials for electrical systems are safe in the various NEC location categories.

Unreasonable Risk—Exists when (1) consumers are not aware that a risk exists; (2) consumers are not able to judge adequately the degree of risk even when they are aware of it; (3) consumers are not able to deal with the risk; and (4) risk could be eliminated at a cost that would price the product out of the market.

Unsafe Act—An act that is not safe for an employee.

Unsafe Behavior—The manner in which people conduct themselves that is unsafe to them or people around them.

V

Vacuum Mentality—Workers think that they work in a vacuum and don't realize that their work affects that of other employees and vice versa.

Vacuums—Pressures below atmospheric level.

Vapor—Substances that evaporate from a liquid or solid. Gasoline fumes are vaporized petroleum.

Volatility—The evaporation capability of a given substance.

W

Waterhammer—A series of loud noises caused by liquid flow suddenly stopping.

Wide Band Noise—Noise that is distributed over a wide range of frequencies.

Willful/Reckless Conduct—Involves intentionally neglecting one's responsibility to exercise reasonable care.

Wind-Chill Factor—Wind or air movement causes the body to sense coldness beyond what a thermometer actually registers as the temperature.

Work Injuries—Injuries that occur while an employee is at work.

Worker Negligence—Condition that exists when an employee fails to take necessary and prudent precautions.

Workers' Compensation—Developed to allow injured employees to be compensated appropriately without having to take their employer to court.

Workplace Accidents—Accidents that occur at an employee's place of work.

Workplace Stress—Human reaction to threatening situations at work or related to the workplace.

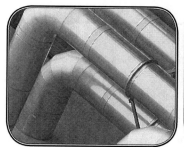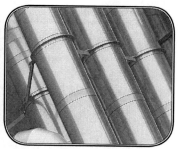

References

1. Breisch, S. L. Hear Today and Hear Tomorrow. *Safety and Health* June 1989, Volume 139.

2. Bureau of Business Practice. *Drugs in the Workplace: Solutions for Business and Industry.* Englewood Cliffs, NJ: Prentice Hall: 1987.

3. Chandler, B. and Huntebrinker, T. Behavior-Based Safety: Multisite Success with Systematic BBS. *Professional Safety*, June 2003.

4. Chemical Manufacturing Association. "Process Safety Management," May 1985.

5. Chemical Week. "Responsible Care," July, 2/9, 2003. Retrieved NEED DATE from www.chemweek.com.

6. Colling, D. A. *Industrial Safety: Management and Technology.* Upper Saddle River, NJ: Prentice Hall, 1990.

7. "Fatality Reports," 2009.

8. Fraser, T. M. *The Worker at Work.* New York: Taylor and Francis, 1989.

9. Goetsch, David L. *Occupational Safety and Health*, 3rd edition. Englewood Cliff, NJ: Prentice Hall, 1999.

10. Hall, R. and Adams, B., editors. *Essentials of Fire Fighting*, 4th edition. Oklahoma State University, 1998.

11. Hammer, Willie and Price, Dennis. *Occupational Safety Management and Engineering*, 5th edition. Englewood Cliffs, NJ: Prentice Hall, 2001.

12. Kavianian, H. R. and Wentz, Jr. C. A. *Industrial Safety.* Reinhold, NY: Van Nostrand, 1990.

13. Ludwig, E. E. "Designing Process Plants to Meet OSHA Standards." *Chemical Engineering*, September 3, 1973.

14. Manuele, Fred A. *On the Practice of Safety.* Reinhold, NY: Van Nostrand, 1993.

15. Bennett, Mindy. "TICs, TIMs, and Terrorists." *Today's Chemist at Work*, April 2003.

16. O'Connor II, J. S. and Querry, Kim. "Heat Stress and Chemical Workers: Minimizing the Risk." *Professional Safety*, American Society of Safety Engineers, 1993.

17. "Plant Security." *Occupational Hazards Magazine*, May 2002.

18. "Site Security Guidelines for the U.S. Chemical Industry." Accessed NEED DATE from www.cl2.com/security guidanceACC.pdf.

19. Smith, S. L. "Combating Stress." *Occupational Hazards*, March 1999.

20. Thygerson, Alton L. *Accidents and Disasters, Causes and Countermeasures*, 2nd edition. Englewood Cliffs, NJ: Prentice Hall, 1977.

21. University of Texas at Austin, Petroleum Extension Service. "Hazardous Waste Emergency Response Training," 1992.

22. U.S. Department of Labor. "Occupational Injuries and Illnesses in the United States by Industry," BLS/Bulletin 2399, April 1992, Washington, DC.

23. "Workplace Violence Awareness and Prevention," U.S. Department of Labor, Occupational Safety and Health Administration, Washington, DC, April 1998. Accessed NEED DATE from www.osha-slc.gov.

24. "Injury and Death Statistics," Injury Facts, 2009. Accessed NEED DATE from www. nsc.org.

25. "Workplace injury, illness and fatality statistics," 2010. Accessed NEED DATE from www.osha.gov/oshstats/index.

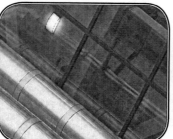
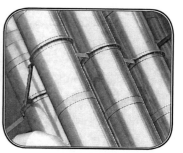

Index